Wisconsin

Dorchester

Thorp

29

Withee

Abbotsford

Unity

Stratford

153

Mosinee

Roth

15

Spencer

Big Eau
Pla ne Res

ek

usta

Loyal

39

12

Fairch ld

MARSHFIELD

10

51

10

NEILLSVILLE

186

Humbird

13

Plover

Alma
Center

95

73

54

Merri lan

Pittsville

WISCONSIN
RAPIDS

12

Hatfield

Dexterville

Port Edwards

39

54

Nekoosa

Wiscons n River

LACK
IVER
ALLS

94

Babcock

173

13

M llston

12

Warrens

Petenwell Flowage

rose

27

80

Arkdale

Roc

21

Tomah

21

FRIENDSHIP

21

SPARTA

Necedah

Adams

16

80

Castle Rock
Flowage

Westfield

5

131

90

94

13

OSSE

New Lisbon

Oxford

33

27

Wilton

71

Kendall

MAUSTON

82

n Valley

Cashton

Ontario

Elroy

16

23

Union Center

12

Briggsville

14

Westby

131

Hillsboro

33

58

Lake Delton

Wisconsin

MUSINGS
of an old country vet

Dr. Paul W. Dettloff, DVM

Other works by
Dr. Paul W. Dettloff, DVM

*Cream Separator Guide – A collector's guide
*Milking Machine Guide – Antique Milking Machines
*How I See It – Carton book
*A Million Miles on Back Roads – Poem book
*Alternative Treatments of Ruminant Animals – 1st edition
Alternative Treatments of Ruminant Animals – 2nd edition
*Out of Print

MUSINGS
of an old country vet
Dr. Paul W. Dettloff, DVM

[signature: Joseph J. Drewes Jr]

Million Mile Press
Arcadia, Wisconsin

Million Mile Press
W20384 Highway 95
Arcadia, WI 54612

Ordering Information: Please contact Million Mile Press.

Printed in the United States of America

ISBN 978-0-9632897-3-5

First Edition

10 9 8 7 6 5 4 3 2 1

Dedication

*This book is dedicated to my wife, Joan,
who over the years has become my love,
my best friend and, in many ways,
a rock to lean on.
You are the best, my dear.
Paul*

MUSINGS
of an old country vet
Dr. Paul W. Dettloff, DVM

A collection of thoughts, musings, poems and cartoons from over 45 years of being a country vet.

On the cover
Lyle Dettloff farm, July 1948,
Grand Meadow, Minnesota.

    This pictures the Dettloffs putting up loose hay with an old grapple hayfork. Paul's father, Lyle (1913-1989), is guiding the hay up into the mow with a three-tine pitch fork. The hay is being pulled up into the mow by Paul's grandfather, Otto (1889-1975), with his team of Belgian work horses. A big hay rope, standard equipment on every farm in that era, ran through the length of the barn out the back and down to the horses.
    Noel (1938) is sitting on the tractor and Paul is watching with Spot, his dog. This entire process became antiquated with the advent of the hay baler in the late 1940s and early 1950s.

Foreword
My 70+ Years Of Dairying — A Timeline

1942 — I was born in a farmhouse outside of Grand Meadow, Minnesota, and was delivered by my great aunt Musetta Dettloff, who was a midwife in the area. I was a normal, healthy boy, and after one good swat from Musetta, I was ready to go.

1952 — As a 10 year-old farm boy, I was milking cows by hand with my father and brother. We had Shorthorns, Guernseys and other crossbreeds, as long as they milked, we milked them. We did not use artificial insemination. Bulls got rotated through the neighborhood as the farmers worked together. I went to a one-room country school, four miles north and two miles west of Grand Meadow. We lived one mile from the school and I rode my bike or walked every day (uphill both ways). I carried a dinner pail and used the outdoor toilet at school, as well as at our farm until 1954. We had no television or extras. I was raised, and worked, in a loving family that taught me the value of a dollar and how to work.

We (my older brother and younger sister) did not feel we were deprived of anything. Chores and farm work were expected. I got two new pair of bib overalls and a new pair of shoes in the fall for school and felt I was living the American dream on our 120-acre farm. Mother had 200 laying hens and on Thursdays we sold eggs as the route truck came for them. We had pigs, feeders and a few sows were farrowed.

We had 15 to 18 milk cows and separated the cream. It was sold for making butter in Grand Meadow. The skim milk was fed to the feeder pigs in wooden troughs. You've never seen excitement until you've seen thirty 40-60 pound feeder pigs come to the wooden trough as you pour skim milk into it. No Nintendo, Sega or Atari will give you a feeling like feeding those pigs.

We made hay loose and put it into the hay mow. Long-stemmed grassy hay was the bulk of our forage. A 12 by 40 foot silo was filled each fall with corn silage. Each cow got about 10 pounds a day in winter and the rest of their feed was hay. On Saturdays we ground corn cob and whole oats with a hammer mill. Each cow got a one-pound Hills Brothers coffee can full on her silage daily. A little bone meal was thrown out, when we thought of it, for minerals. That was our ration in 1952.

I remember being at a neighbor's filling silo in the early 50s. There were some heifers and a couple of steers in the barnyard chewing on the boards of the fence when a mineral salesman drove in. The owner of the farm, Henry Miller, was a big man. He was a strong, physical person as well as strong-willed. The mineral salesman (minerals were just being introduced in southern Minnesota) asked Henry what he did when the cattle ate up the whole board, and Henry continued with what he was doing and replied that he just threw in more wood. With that, the salesman left. That was the conservative mindset I was raised in.

We rarely called a veterinarian for anything. My dad used Piperazine wormer. He used a razor blade in a potato for castrating hogs and our main veterinary items were pine tar and Lysol. Animals rarely got sick as nothing was pushed much and they got colloidal minerals from our still-balanced farm.

We had no sprayer. Corn was planted with a two-row John Deere planter. We rotary hoed once and cultivated three times, weather permitting. My dad did put lime on everything. We had a short rotation, grazed in the summer, and followed the Milwaukee Braves religiously.

1962 — Found me in college at the University of Minnesota in the pre-vet curriculum. I had two jobs most of the time until veterinary school, then I could only handle one as the time demands for school were too great. I went through seven years of college and never borrowed a penny from anyone. Most of our class did the same. Forty-nine of us, forty-eight males and one female, graduated in 1967. I had a net worth of less than $100. I bought a house and veterinary practice in Arcadia, Wisconsin, in 1967, with the help of an uncle who co-signed the loan, and I hit the ground running. In 1968, I bought a new Dodge pickup with a four-speed tranny for $2800. The call charge was $5, and I would treat a Milk Fever case for a ten-dollar bill. Things were great in the dairy industry.

1972 — In 1972, we had a multiple-man practice with three owners and a fourth veterinarian employed. We owned a new clinic that we built in downtown Arcadia. New drugs were coming out monthly. Silos were going up weekly in nearly every valley and on every ridge. Every farm in Trempealeau County had dairy cows on it. Preventative medicine and herd health programs were now the normal thing to do. We started treating all dry cows with the new antibiotic-laden dry cow tubes. The new multivalent three-way vaccines would take care of the respiratory problems. When these new innovations hit in the 1970s, there was no slaughter or milk withholding. Then people became sensitive to penicillin, so we had to put withholding times on the antibiotics. There were no cell counts back when I started. If it went through the strainer, you sold it.

Herds were getting bigger as one neighbor bought out the old fellow next door. The benchmark to make a living went up to 45-50 cows. The thought was, "Let's quit pasturing, you can raise more forage on those hills. Dry lot your cows, go to corn, alfalfa and soybeans. Get that production up!" Bigger and more silos went up. A.O. Smith had a sales force and program like we had never seen in agriculture. Flail choppers, big chopper boxes, and the 40 horsepower tractor became obsolete. One hundred fifty horsepower became commonplace. Production Credit Association was big. PCA and the ag bankers coined the phrase "cash flow." Everybody during the 1970s made money if they did half a job. The technology was wonderful. Drugs, sprays, insecticides, herbicides – you name it, if you didn't follow along, you were against progress. We all followed along with the pack, as we hadn't seen the side effects yet. They were building as the technology pendulum was swinging.

1982 — The late 1970s were a rough spot in my life, as I lost my first wife in an auto accident and lost my oldest son at six years of age to leukemia. I left practice from 1978 to 1982 and was in industry. I met and married a wonderful person and we had more children. In 1982, I returned to practice as a solo veterinarian at Arcadia, Wisconsin, where I had started.

Farmers were starting to leave as the prices were not keeping up with the expenses. Attrition was taking out the older ones and the younger ones were not as eager to jump into dairying as they were earlier. Herds were getting bigger,

and bunker silos were what was needed for more tonnage. Some of the antibiotics did not continue to work as they had before. We started to see tough mastitis cows that had organisms, when cultured, that were now resistant to antibiotics.

We still had new drugs, vaccines and antibiotics coming out. They were much more expensive because of the research and development required to create them. Prostaglandins and hormones became more sophisticated. Nutritionists pushed the production envelope more and more in order to get more milk. Meat and bone meal, animal fat and blood meal were thrown into the mix in the least-cost rations. Corn silage for energy was increased. The level on minerals and vitamins was raised as production went up. Displaced abomasums became the gauge of a veterinary practice. They were commonplace. Johne's disease was showing up more and more. Then we began seeing it in heifers.

My first deposition for a stray voltage lawsuit took place in the early 1980s. Six more were to follow for me. Stray voltage was a new aspect of veterinary medicine. The little country milk co-ops were merging to stay afloat financially. They thought they could be more efficient if they were large. Many of the drugs that I had started practice with were now illegal. DES (Diethylstilbesterol) was the first hormone debacle that skipped a generation to hit the daughters of mothers that used it. DES was used in veterinary medicine in large quantities. Chloramphenicol, an antibiotic we discovered, could cause aplastic anemia in humans and shut down their bone marrow. Penstrep, in oil, stayed in the system for long periods of time. Using the sulfas in lactating cows was outlawed. Drug companies were merging and being bought up. European companies were buying

our U.S. drug companies also. Veterinary practice was heavily associated with drugs, drug dosage and high production.

The veterinary and dairy industries were suffering from tunnel vision and were relying on lots of external inputs for high production. Dairy farmers and veterinarians were actually the pawns of technology and industry.

Two things happened, and we did not see them happening as the changes were so slow. It was like watching your child grow up. All of a sudden, he is taller than you, graduates from school and leaves home. Technology did a number on our soils. We killed soil life. One spoonful of soil should have two billion living organisms in it – bacteria, fungi, nematodes and protozoa, to name a few. These organisms recycle everything, build organic matter, absorb moisture, give soil tilth and help restore and balance trace minerals. All the different –*cides* we put on the soils, plants and fields, killed our helpful little friends out there as they are all made of cells. The second major dairy problem created in the 70s and 80s was acidosis. We forgot the rumen was made for forage – that's grass and hay – long-stemmed, nutritious grass and hay. We found a few seeds would kick the energy and give a spike in production. The pendulum swung way over on the seed side with the emphasis on high production. The side effects set in: bad feet, lowered immune systems, poor breeding, laminitis and poor colostrums.

1992 — By the early 90s, we had lost the lion's share of the family farms. We were getting the mega-dairy setups. The little farms with no debt or little debt were biding their time, making it because the wife worked in town and got the health insurance and a 401K for herself. When these farm systems get old, they can't be replaced, as the capital outlay for the land and machinery and cattle won't begin to cash flow as the debt is too high.

I saw my first organic farm in the early 1990s. They were producing organic milk because some people did not want to take all those free radicals and chemicals into their systems. My two sons learned how to eat properly from their wrestling programs. They drank juice, ate fruit, and told me about aspertame. My two college graduate daughters are knowledgeable about organic food in the Twin Cities. All of a sudden, we have a generation of young non-farm people (98 percent of all people) who want to know and are questioning what is in their food. My organic farmer turns into two more. These people are worried about their earthworms and soil. I had nothing to offer them except saline and glucose. My organic trip began. I learned just as the whole movement did, from the ground up. My clients needed tools and help and so my quest began.

2002 — I see great changes coming in the dairy and veterinary world. There are now some of the big, high production, high technology, high input dairy operations going bankrupt or dissolving their assets as the ten dollar milk will not compensate their cash flow outlay. I am seeing the middle group of small farms, with the older operators, being phased out by attrition as no one can capitalize the land, cattle and machinery or even begin to cash flow a dairy operation in the colder part of the United States. The third area of dairying is the certified organic and the biological group, where they have transitioned out of conventional dairy operations. I have 15 certified organic dairy farmers in my practice and six more that will be on the organic truck by spring 2003. I see the veterinary profes-sion slowly becoming aware of the organic movement. The younger female veterinarians are the ones that are showing the most interest.

The strength for the organic movement is coming from the younger generation who wants to know what is in their food. I see the dairy industry gravitating to two poles: the big, high-input, usually acidotic setups and the small organic family farm with low inputs. The factor that will dictate who will be around in the future will be determined by which one is sustainable over the long term. The organic one will have the edge as they are soil and environment friendly. Another group that is growing and will be a factor in the future are the seasonal calving grazers. They calf from March to early June, take peak of summer pasture and dry up in December and January. These are quite soil and environmentally friendly and are low input. This group makes sense and will continue to grow, some being organic and some not.

Another change that helped the organic veterinarian and organic dairy farmer is the USDA setting up the National Organic Standards Board (NOSB) and the National Organic Program (NOP). The NOSB has worked hard and long to develop a national list of safe ingredients for the organic dairy farmer to use. This was finalized in its initial stages in October, 2002.

2012 — Finds me out of veterinary practice as I quit taking calls on June 30, 2002, as I needed a new hip.

After a million palpations I wore out my right hip, as I used my left hand for rectals, wearing out my femoral head. I became the staff veterinarian for Organic Valley Co-op at La Farge, Wisconsin, and consulted with Lancaster Ag Products in Pennsylvania, owned by my good Amish friend, Reuben Stoltzfus. I have traveled exten-sively the last 10 years. Organic Valley in ten years went from 411 farms to 1500 dairy farms. Lancaster Ag has another 500.

I put on barn meetings, write articles, do pasture walks, and educate farmers and others from the soil up. Dr. Paul's Lab, which my wife so dutifully manages, has also grown. We manufacture organic animal treatments, tinctures, boluses, botanicals. Basically, there is a natural treatment for everything. I have had the benefit of growing up and practicing in the James Herriot paradigm. I am sad to report those days are over. What I did my last 70 years is gone. We are either mega dairies or organic.

When I started practice in the '60s every driveway had Ma and Pa and the kids eking out a living on the farm with a garden and a biodiverse farm, just one step away from subsistence agriculture.

The cartoons, poems and anecdotes are from my section of life that will never repeat. I have totally enjoyed the ride with my clients and wonderful growing family. All of you go forward with a positive attitude, knowing that attitude is 90% and circumstances are 10%.

Hope to write again in 2022 with 80 years of dairying. If not – enjoy.
Dr. Paul

Take the contents of this book for humor's sake only.
In no way do I intend to offend any group or person. Life is short and easier with humor.
I hope this puts a smile on your face — just for one day of your life.

As you'll notice, there is no bibliography in this book because all the thoughts, drawings and ramblings come from the mind of the author, Dr. Paul W. Dettloff, DVM.

Collecting

I grew up using and washing out cream separators, and started collecting them when I started practice and saw an unused separator standing in the corner of a building on most farms.

After years of collecting, I had two pole buildings full of them and wrote a guide book for other collectors. I was always on the lookout for rare ones. Quite a few I was able to trade with the farm client directly. I also watched the farm auctions, as separators appeared commonly in auctions in the 60s and 70s, as they were no longer being used. They had about a 50 year span from the early 1900s to about 1955 when everybody went to a bulk tank and sold whole milk.

The cream separator took dairy farming out of subsistence agriculture to marketing a product for the masses every week. It totally changed diary farming.

I saw this one auction in the paper that listed a nice, clean, big separator. What I would do, as I didn't have time to go stand at an auction all day, is I would stop in before the auction, look at it and then call the clerk of the auction and tell him how high I was willing to go for the item. I was personal friends with all the clerks as I tested cows with them for brucellosis before the farm auctions.

I had three clients that were all brothers, that lived next to this one farm that was having the auction with the big, heavy separator. It so happened that they helped the neighbor line things up the week before the sale. The oldest brother, Dan, had the job of getting the items out of the upstairs of the granary. Yup, you guessed it - that's where that big, beautiful separator was. After a lot of grunting, lifting and sweating, they managed to get that beast down the wobbly non-OSHA steps, out onto the grass and into the machinery line-up. The brothers stand and stretch to get the pain out of their backs and the oldest says to his brothers, "Wow. I'm glad that's done. I suppose the next guy to lift that monster will be Dettloff!"

Well, the auction comes, and I had told the clerk that if I ended up owning the separator, if he could just ask my good client, Dan, to haul it to his place for a few days as I was leaving on vacation the next day.

Sure enough, I get my prize with the winning bid. As everybody is leaving the clerk stops the three brothers and informs them of my request to have them haul it over to Dan's farm and keep it for the good old vet.

It was nice to have super clients like that!

If Cows Could Talk

If a cow could talk, it would make dairying easy.
They could take roll call of who's sick and queasy.

A list could be made and put on the bulk tank,
"Check out the big heifer, that sore on her flank.

Big Red is limping with foot-rot disease.
Check the lungs on Molly's calf, he's starting to
  sneeze.

I'm going to freshen tonight. It feels like a bull,
check on me at midnight, I might need you to pull.

Do something about that heifer with the sharp horn,
We can't eat hay at the bunk, she hogs all the corn.

Hey! I'm done milking, get that thing off my udder,
Get off the phone or I'll kick it in the gutter.

I just ate a nail, it hurts and I'm sick,
Get a magnet and make it real quick.

My foot is so sore I can't walk and graze,
Get the hoof knife and rope, my leg I will raise.

Today I'm in heat, with you it's a guess.
Call in that gentle technician from ABS.

Hey, this pen is wet, the bedding's all soggy.
Clean this place up or we'll step on your new doggy.

My right front quarter is sore, I'm sick as can be,
Go to stripping it out and give me an I.V.

For milking you're late, and we're starting to
  prance,
You were out too late at that stupid wedding dance.

That hired man you had for help last summer,
Don't hire him back, when you're gone he's a
  bummer!"

It's good they can't talk, it's best this way, you bet!
Cause I know every morning they'd be calling the
  vet.

They'd want a vacation and maternity leave, too –
Yes, I think it's best they only can "Moo."

1

Roots

In the 1800s a farm family of peasants was toiling
   away,
They threw down their hoes and said "Let's go to the
   USA."

His name was Charles (Carl) A. Dettloff and his
   wife was Mary Dorman
The seas were high, the winds had the seas a
   stormin'.

Their six Prussian born children, with all their
   belongings in one big crate,
set off on a ship for Minnesota, a Midwestern farm
   state.

They were hardy travelers as they said goodbye to
   the Rhine –
Yes, I came from a very strong homo sapien line.

With an ox cart, Durham cattle and some seeds of
   wheat,
They started walking to Grand Meadow, Minnesota,
To get some good, black soil under their feet.

The Homestead Act gave them 160 acres to settle
   and till.
You see, we took it from the Native Americans
   against their will.

Typhoid, Cholera, the flu of 1918, all took its toll in
   time
But not the Dettloffs; I came from a hardy homo
   sapien line.

Oh, we weren't perfect, we had one ancestor whose
   genes flipped off its tracks.
He did his in-laws in one day with a firewood axe.

Mostly farmers that worked with the soils to feed the
   masses,
They spread all over America on many states'
   grasses.

My great-grandfather, born in 1850, was Charles'
   second son.
He married a German gal named Rieka Eilers just
   for fun.

They had eleven children, ten lived just fine.
My grandfather was named Otto Reinhart, who lived
   on the county line.

My grandfather's brother, Walter, spent WW II
   fighting on the Rhine.
Yes, I came from a mighty strong homo sapien line.

He married a strong woman named Leona
   Biederbeck.
During the depression he, financially, almost lost his
   neck.

They hunkered down, quit spending, planted a
   garden, raised animals and lived off the lands.
They held on and survived with persistence, using
   their mind, back and hands.

They brought forth a son, Lyle, and a daughter,
   Agnes.
You see, after World War II we worked out of our
   economic mess.

Farming was good, we were growing and prospering
   in the USA.
We worked hard as a society and had time to play.

The Dettloffs – yes, they were doing just fine
Yes, I came from a mighty good homo sapien line.

I went to college and studied to become a large
   animal vet.
It was very good to me, oh yes, you bet.

Now we have our country in a financial mess.
How we are ever going to pay off our debt is
   anyone's guess.

Oil prices are going up. We may run short of food
   and water is short.
Oh, it's terrible to listen to the news to hear the bad
   report.

I smile and think about my ancestors' previous
   troubles and worries so sublime.
I get reassured that we will survive as I came from a
   very strong homo sapien line.

Dr. Paul Dettloff, DVM
6-14-2010

I'm Back

If you have been waiting for another book on the
    edge of your chair,
Well, sit back, here it is, only this time it's coming
    with grey hair.

I've a Titanium hip, a zipper on my chest after heart
    repair,
Grandchildren, that Grandma and I do so care.

I've worn out many vehicles, saved some cows and
    sent some to slaughter,
Our cute little ones in our fold call me Gran-fodder.

Now I'm speaking of sustainable ag, something I
    never planned,
In March I'm going over seas to speak in New
    Zealand.

Close relatives have died, I've had to probate two
    estates.
I'm personally stepping closer to those pearly gates.

I wrote a book on alternative remedies and
    sustainable tools
So organic farmers could treat sick animals and
    abide by the NOP's rules.

I've traveled coast to coast and thrice over seas
To help dairy men address every lameness, sickness
    and wheeze.

So, in this book I've got short stories, poems and
    cartoons
So the next generation can see what I did on early
    morning calls, midnight emergencies and chute job
    in afternoons.

I just love it when the vet's already been here. Everything is ripe and ready.

Too bad about Roger being allergic to penicillin.

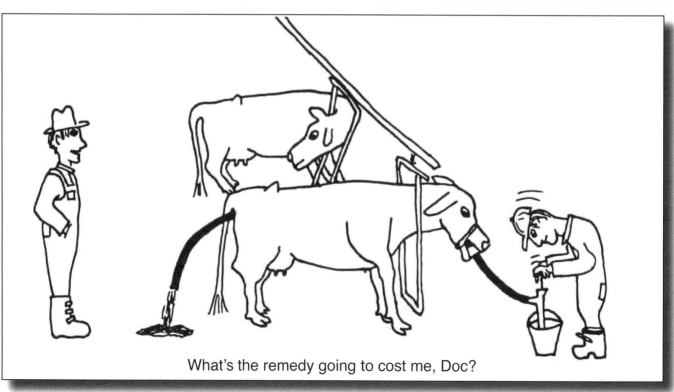

What's the remedy going to cost me, Doc?

Have you had this place checked for stray voltage lately, Clyde?

Parasites

There once was a pig by the name of Roy.
He was afflicted with every Parasite nature could
   employ.

He had a little mite, that itches so strange,
The Swine Producers, they all call it Mange.

Then he picked up Round Worms, when just a kid;
The long, round, ugly old Ascarid.

In his bed where he slept, curling up to his knees,
There was no rest; it was infected with Fleas.

Something was biting; it sure wasn't nice.
Poor little Roy was loaded with Lice.

Life was an itch, scratch, coughing for air,
Parasites were all over – stomach, ears and hair.

To get rid of them, his life would be finer,
His muscles were sore, from encysted trichiner.

Life was so miserable, he'd die, what the heck!
Then along came relief, in the form of Ivomec.

One c.c. sub-q, was more than he needed;
But he was in tough shape, the farmer conceded.

A short time later the Parasites jumped ship;
This Ivomec was destroying them with every sip.

Now Roy is happy, healthy, and lean;
His hair is all shiny, his bed is all clean.

There's not one Parasite, not even a spec;
Thanks again to Merck's subcutaneous Ivomec!

Clyde, I think you should know that one of these cows is lying to you about her butterfat and protein tests.

I know earthworms are good for the soil but this is the second heifer calf I've lost this week.

Before I check out your cow, Clyde, you say you want to show me your new website?

So, you just want the three house cats vaccinated.

Junk

Being the father of the farm and household makes
  me a responsible hunk.
That means I'm the party that disposes of all the
  junk.

Having a big family the volume of debris is huge; we
  have so much waste.
So I've become a sorter, a burner, a squasher and
  the toxins I have encased.

Years ago we solved it all by having a dump on the
  farm,
Now with our technology and chemicals, we find
  items that can do us great harm.

If you're not careful how you dispose, it will get you,
  no matter.
It will return into your system thru the food chain
  and water.

When I moved onto my farm in nineteen and seventy-
  four,
I hauled the farm dump away to clean up the eye
  sore.

But, where did I go with the mess of metal and
  glass?
I went to the bigger dump that the township ran,
  supposedly first class.

Now that has been closed, it seems stuff has been
  seeping away.
They received an ultimatum to lock the gate from the
  EPA.

So now we sort and wash and crush, oh what fun,
And it goes to a clay-lined landfill and we pay by the
  ton.

I've got a big barrel for the aluminum cans we
  squash like a toad,
The pretty labeled cans the advertisements sell to
  lighten life's load.

The glass gets delabeled and washed, the green,
  amber and clear,
We take it to a building that says "Recycle here."

Now, I'm old fashioned 'cause I have a barrel to my
  name.
The cardboard, paper and all combustibles
  disappear in a flame.

Only problem left is the ashes, black and sooty as a
  witch.
I just pack then into some little erosion ditch.

Then I'm left with a barrel that's rusty and eroding
  away.
I'll throw that on with the cast iron and give it to the
  junkyard one day.

Major problem with the junkyard and my lovely
  wife;
It does cause us a bit of marital strife.

I leave with a pickup half full of scrap iron and tin
But, when I come home, I've got the pickup loaded
  with neat old cast iron right up to the rim.

What'd ya pay for this piece of junk?

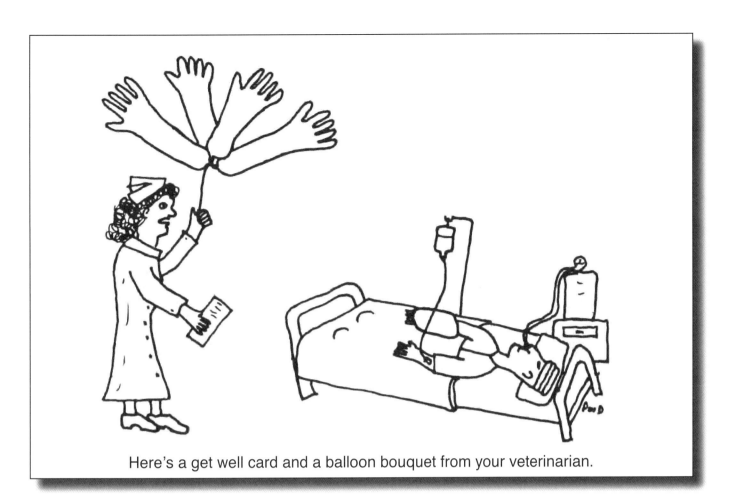

Here's a get well card and a balloon bouquet from your veterinarian.

Clyde, when did you start using a Mohammed bull?

No more dead BGH cows for you, Mildred.
How are you ever going to fly with those things a'hangin?

Do you realize, Buzzy, that there are 40 percent less carbs in a pig than in any ruminant?

Communing With The Elements

It's Christmas Day morning, December 25, 2000. I'm sitting on a ridge top in Trempealeau County, Wisconsin. The stars tell me it will be clear today, and cold. It's 6:25 a.m., just before dawn, and it's just me and the solar system.

I rolled out a little after 5 a.m. to greet a womb pushed out on a bovine. She had calved early. The calf was in the gutter with the womb pushed out on top of the poor calf. The cow was down. Luck was with me as she got up, the calf was revived and the womb wasn't badly swollen yet. So with spinal and warm water, it was replaced.

I will retrace my trail home over 15 miles of ridges and valleys. I've not met a car yet. The only things stirring are diary farmers and one happy veterinarian.

The trees are now covered and stately this morning. The deer are bedded down and the crows are in their rookeries. I get a feeling of quiescence and serenity.

I can imagine a Paleo man 10,000 years ago, standing on this very same spot, going home with an animal he harvested slung over his shoulder to feed his kin and clan. I bet he stood here in this exact spot with my same feeling of all being right with his world, before he descended his trail down the valley. It makes me feel so small in the realm of time. My impact on this planet in the next 10,000 years will be like a grain of sand on our shores.

I must crawl back into reality and treat one more sick animal before I go home to my Holiday and family. Everyone needs to stop on a ridge and reflect past our present day strife and just appreciate where we are in life.

Well, Clyde, after an extensive autopsy and my seven years of college training using scientific analysis, I would say there's a 75 percent chance that your animal died from Black Leg.

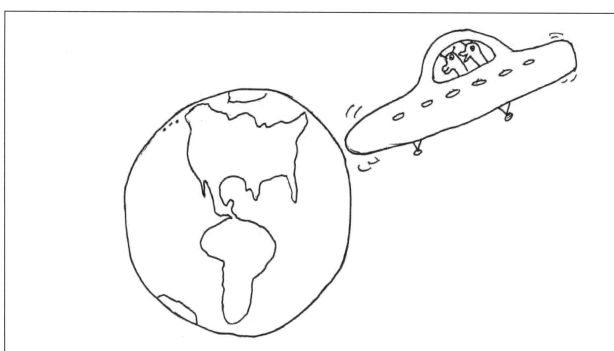

Well, the homosapiens are gone. The Genetic Engineers mixed up fertilization with defecation, and ovulation with urination, and they all died in the bathroom reproducing.

I think you're running a little too much vacuum there, Clyde.

Ugg, please stop playing with daddy's wrenches.

Mail Boxes

Everyday, five hundred mailboxes whiz by my
  windshield;
Some sticking out into the road, some way back in the
  field.

The most common is the one-post job;
Just nailed on top, sorta like a knob.

I enjoy the ones setting on some old farm culture;
Some have their door open, looking like a beak of a
  vulture.

I see quite a few using a ten-gallon milk can;
They stand with their shoulders, just like a little man.

There's one down the road, just past a bump,
Whose box is welded on an old cast iron pump.

I see old gas engines, whose pistons did fail,
Out by the road, waiting for mail.

An old iron wheel is part of the mailbox club;
It never will turn again, the spokes nor the hub.

The wooden contraptions that decorate the ditch,
Some look like they were ridden by a witch.

An ingenious one, rigged up with a spring,
When the snowplow hits it, away it will swing.

Another unique one, it took a creator,
Is perched on top of an old cream separator.

There's Uncle Sam, painted red, white and blue,
Holding up the box, pointing "I Want You."

There's a rear end used and, yes, a drive shaft;
To weld them just right, it all took some craft.

You can see all kinds, some neat, some a mess.
We all complain about the US mail, but it's better than
  Pony Express!

13

I think you can back off on your herbal grub and louse treatment, Clyde. You seem you be getting a little overgrowth.

Been here long?

A Bad Day

I was driving down the road eating a bar of candy,
Listening to country music, a good song by Moe
Bandy,

When in the window flew a bumblebee – made a
loop and did disappear.
He's gone, there's nothing. Three miles later he
stung me on the ear.

Next stop was a big, tall heifer. She was tranquil,
quiet and shy.
While giving her a physical, her left leg landed
squarely on my thigh.

Back in the vehicle, turned the radio up and open a
Diet Coke,
This time it's Conway Twitty expounding on his heart
that's been broke.

The very next call, a beef calf with an infected sack;
Lariat on the neck, a wrap around the post, and
thump, I'm laying on my back.

I stagger to the vehicle, turn on the radio, and, wow!
Willie Nelson's telling me it's too late to do anything
about it now.

I stop at an intersection, someone yells 'That cow
you treated finally died.'

For relief, I change stations and pick up good old
Charlie Pride.

Next stop is a breeze, just a simple ordinary cow off
feed;
Except, when I left, guess where I laid my nose lead?

Back to the radio, to lift up my spirits and temper my
fright,
Just to hear Paul Harvey tell Columbia has more
druggers to extradite.

Last call this sunny day, without any thunder or rain,
A big Holstein cow stepped on my toe; you talk
about pain!

Stumble into the seat, turn up the volume, hit the gas
and spin a tire;
Good old Johnny Cash, with trumpets and all, has
himself a ring of fire.

So, at last, I'm caught up and gladly home I arrive,
Turn off the radio, stomach empty; glad I'm still
alive.

"How was your day?" true concern from my wife
and honey.
"Oh, good. Quite a few calls, and I made us a little
money."

Yep, Clyde, Doc always said he hates these alert downers.

15

I said three days of penicillin, not three moons.

They never did figure out the wheel principle, did they?

The Record

When I was a kid and first went to a little one room
    country school
I was not perfect; I did bend and sometimes break a
    rule.

But I would straighten right up when I heard a
    teacher say,
"This action might get recorded on your record
    today."

If someone was really bad, caused tension, unrest
    and general discord,
Then that might just go down in the "Permanent
    Record."

That was the ultimate fear of kids, to have that
    written a terrible strife,
'Cause that would follow you wherever you go the
    rest of your life.

Your college would see it, and of course the Army
    and Uncle Sam.
Your life would be scarred, you'd be in a terrible jam.

Your potential employer and the new boss would be
    especially mean
As they would get a copy, over which they would
    gloat and glean.

Whatever happened to the "Permanent Record" they
    kept on that long list?
Did it go into a computer and someone erased the
    disk?

My kids have no fear of such a list or error they
    make.
They just laugh and say you were too dumb to know
    it was a fake.

Even if it's on a computer now with a list of bad
    habits or any sad feat,
They are not worried because in six years that
    computer will be obsolete.

But I keep telling them to shape up and keep the
    permanent record clean.
When they get older with kids of their own they'll
    know what I mean.

It really never was written down in a book or on a
    wall or ceiling.
It's the look on a person, his actions, character, and
    that unsung feeling.

Doc, cheers to 40 years of dairy practice without ever using a sleeve.

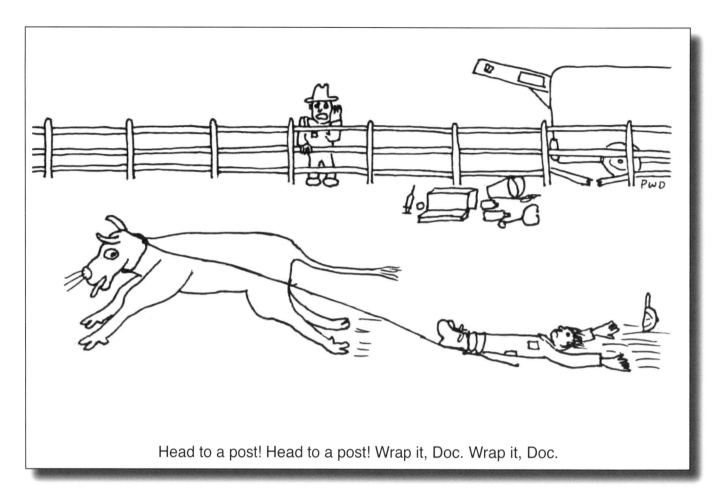

Head to a post! Head to a post! Wrap it, Doc. Wrap it, Doc.

Global Warming 15000 BC

Tonto!! Would you please stop puffing on that stupid peace pipe
or you'll melt our glacier and then what are we going to do?

Hardware

I am writing about hardware because I was born at the right time to see the dairy world cycle through some interesting phases.

When I hit practice in 1967, our cows were still pretty much on permanent pasture in the summer and dry hay in the winter. When we let cows out in the spring, they would always eat dirt due to lack of mineral feeding or not enough salt over winter. You would see about four weeks of hardware also.

Then in the 70's everybody started green chopping with flail choppers to open up the hayfields for harvesting. Flail choppers sucked everything up off the ground. In the 70's a new silo went up every four days in my area. A. O. Smith had the best sales force in America with the farm credit man sitting in the front seat ready to finance them. Man, then we really started seeing hardware cows.

At that time I was senior man in a four man veterinary practice. Every day we would talk about who had another cow with hardware. Then two things happened. Sperry New Holland put a magnet on their chopper and it would stop the machine dead when metal was in the machine and A. O. Smith put a big $90 magnet on their unloader. The hardwares dropped significantly.

We had become experts on hardware. We learned that all magnets don't go down all the way – some stopped in the esophagus half way down the neck where the food tube goes from ventral to the left side. If left there, they can regurgitate them. We followed every magnet with a little oil filled compass. They had to be on the bottom left side behind the left elbow. A few would land on the fold between the rumen and reticulum on the rumeno-reticulo fold and not work. Every cow that received a physical got compassed to see if she had a magnet and if it was in the right place.

A Muffley Magnet Retriever was bought from an old veterinarian, Dr. Muffley in Lewisburg, PA, and we became experts in its use. This device was a long black plastic 1¼ inch tube with a flexible end on a string to pull the end back towards you with a big magnet on a cable that you would push into the reticulum. It took some expertise and practice, but once you got the hang of it, you would grab the wire filled magnet and pull it all out. You went down again as once in a while a piece would come off in the throat. The most I ever got was 28 pieces of fencing wire about ½ inches long on one magnet. These little ½ inch pieces are just perfect to fit into the honey combed reticulum. If we had a fine chopped haulage farm, a second magnet would be given as the little ½ inch pieces would stick out on the end of the magnet like a prickly pear. When we started retrieving them, we figured out the pieces weren't long enough to attract both poles of the magnet, so we would give two magnets and then the pieces would lay in the groove between them.

When we started retrieving and our farm clients got to know that there was an alternative to getting rid of the cow, we went back on a lot more hardwares a second time and either gave a second magnet on haylage or retrieved them. There were a few clients that would call back even if the cow got better, as they would want the magnet retrieved, as they were curious as to what was on the magnet.

Now hardware is not nearly as common, as people are more aware of metal in the feed.

A second thing that is happening is very few of the instructors in the veterinary schools have had years of practice and it takes a while to be able to feel a hardware, so our young graduates have an educational void.

I do a lot of phone consulting, and I'm getting calls saying, "My veterinarian has seen this cow three times. It's not Ketosis, the uterus is fine, udder is fine. They can't find a DA and she still won't eat." I'll ask what her temperature is and it will usually be 102.8 to 103.2 degrees. I'll then ask if the veterinarian uses a compass and the usual answer is "No." I'll tell them to give her a wet magnet or put soap on it and give it to her.

Clyde, you did tell me those Neanderthal were mixed up and wouldn't get out of the ice age.

Hardware is not a disease. It is caused by what the cow eats. Her rumen is the cause. The cow has what is known as four stomachs. Actually, she has a true acid producing stomach that is shaped very similar to the human stomach, and it secretes acid just like ours. That is the abomasums and it is called the fourth stomach. The other three are actually dilations of the esophagus. The big rumen is the fermentation vat that the microbes are in to digest grasses and forages. The bottom front of the rumen has an out pocketing that is honeycomb lined. That is where the grain goes and the heavy stuff like metal, wires, nails, staples, broken gears, and anything else small, sharp and pointed, lands.

The next stomach, the omasum, has the function of removing the rumen juices from the food before it goes into the abomasums. The reason this is so serious, is that the reticulum lays next to the heart, only separated by the very thin diaphragm. The rumen contracts two times a minute, pushing metal forward into those honeycombed partitions, right towards the heart.

Why do cows eat chunks of wire and metal? She eats huge mouthfuls of grass or hay and swallows them for chewing later. When she is done grazing or eating, she will go lay down and bring up a cud, which is about the size of a rolled up wash rag. The cow then chews it up into little pieces and swallows it for the microbes to break down.

When a cow picks up a piece of metal, she will immediately spike a 103 degree temperature, her rumen will stop, and she will silently show a little pain in her eye. They will lay down a lot and don't want to get up or they will stand a lot and not want to move. Sometimes they will have a noticeable heart beat when you stethoscope the rumen.

It takes a few years of practice to really get into diagnosing what you see. Historically speaking, in the 70s if you couldn't come up with a definitive diagnosis and there was a little temperature, it got called hardware. After thousands of physicals you can just sense a hardware.

Aluminum is a different story. Mountain Dew cans won't stop a chopper or get hung up on a magnet. They are deadly. They come in chunks about the size of an old silver dollar with very sharp edges. They cut like a knife and will kill a cow in 48 hours. A second aluminum problem in my Western Wisconsin, deer filled hills, is aluminum arrows from the bow hunters. You get a 2 inch piece of an arrow shaft in the hay or haylage and you have a deadly weapon.

Clyde, at last…

…a Sharples!

Well, I'll be darned. Here's her problem. Hardware Hank.

So Doctor, I take it you do mostly large animal work.

Now I know why you wanted a vet and not a barber to trim the hare off your neck.

or

No, we won't need an M.D. for this operation as long as I keep my incision on the rabbit.

You know what? They taste a lot like chicken.

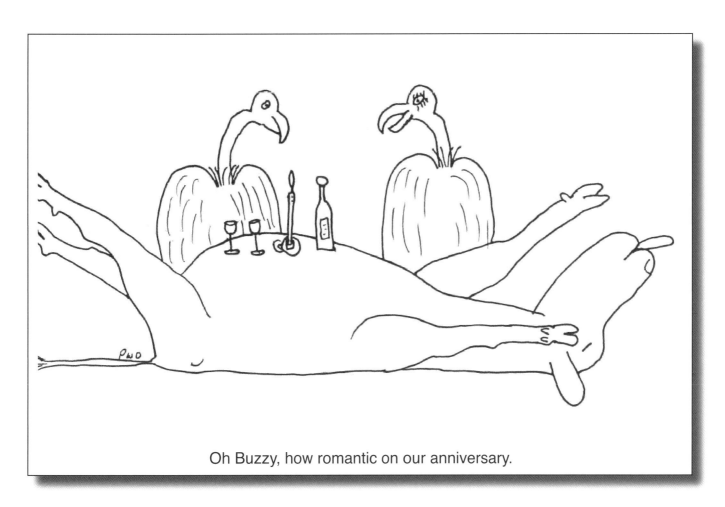

Oh Buzzy, how romantic on our anniversary.

Hardware!

If Separators Could Talk

I've layed in this old granary over forty-six years;
Imprisoned in this dark corner, I've shed many tears.

A burlap bag on my legs usually filled with mice;
Oh, when I was young, my paint was so nice.

One day, the door opened. A man checked me from
my toes to my head.
"Yep," he said, "I'll add it to my collection in my
pole shed.

"Don't scratch it," he said. Hey, this guy's okay;
He loaded me up and hauled me away.

He carefully set me on a board in a garage so nice;
No more dark corners, smelly old burlap or mice.

It tickled and felt good, as he washed me with
gasoline;
He cleaned all my gears up, now I'm a running
machine.

The pains he took cleaning me, he's a real pal;
Why, he spent one whole evening just on my decal.

The rust and the dust were removed from my spouts
and my tank;
A new set screw in my handle, to fix up my crank.

He rubbed me down with WD-40 and oil – I liked it
so much;
I remember him saying it was the finishing touch.

Then on a cart I was loaded, and up to his shed;
He set me next to a Belgium Melotte, that's shiny
and red.

It's international, the group – Swede named Domo, a
Finn called New Prima;
The one to my left is a little DeLaval Number 2, I
call her Katrina.

We rest in the night and talk about our working
years all day;
I'm getting sick of that Canadian Renfrew, all he
says is "Ay."

Now, we're a happy group and about once a week;
Our owner comes with some strangers to take a
good peek.

A lot of our ancestors were junked or melted during
the war;
We are really thankful as we know we'll be preserved
ever more.

Separator Events

I'm an old Cream Separator, cast in 1901.
When I was young, my bowl really spun.

I had a long good life; got cranked morning and
nights,
But I was replaced when the farm got electric lights.

To the granary I went, into a corner that was dark;
No cranking or cleaning, my existence was stark.

Then along came a man in a pickup with a bug
reflector;
Told my owner he was some kind of separator
collector.

He actually bought me, gave my owner a check,
I got a new lease on life, I was happy, by heck!

I was elated, for now I could fulfill a dream –
Someday I'll actually produce more skim milk and
cream.

But my new job isn't to work, as now I am on display;
It's a super retirement, no work, all the benefits, I
just shine away.

I'm cleaned up like new, my brass plate with patent
dates and name of inventor.
I'm also excited, next fall I'm going to Northwest
Ohio to Liberty Center.

All the Separator Collectors, big and small, rich or
poor, they are all worthy of mention,
Will be gathering at Tom Mitchell's for the Fourth
Annual Separators Convention.

The second weekend of September is a date to be
noted,
They will talk, trade, spin yarns and eat until they
are bloated.

So, I will just stand here, a shining with one big smile;
Cause I'm going to be shown next fall in Northwest
Ohio.

Rookeries

Have you ever seen a crow convention?

I have twice in my life.

In 1974, I bought a farm and moved to the country. In about 1976, I stopped in during calls to pick up some drugs and the ridge on the north side of my farm was alive with crows. Not 20, 30 or 40 crows – there were over 1,000 crows. They were swirling, circling, cawing on the entire side hill (oak woods) and ridge.

I dropped everything and started walking towards them. This was before 4-wheelers. I got to the bottom of the woods and sat down, just to watch. They sort of quieted down when I approached, but within a few minutes they were right back to their crow antics. I watched them for probably 20 minutes. I had never seen so many crows acting so goofy, like they were showing off their flying skills.

I told a few people about this. They listened, but questioned me as to how many crows I saw.

One day, a few years later, it was explained to me by a very wise gentleman. He told me he was totally aware of this crow phenomenon and explained to me what they were doing. He explained that it was to get genetic variation and to prevent inbreeding within each little rookery. He explained that this happens someplace every 10 years or so. What triggers it, he didn't know. About 1996, I was on one of my client's farms and he told me the same story, that he had witnessed over 1,000 crows carrying on on his farm the day before. The next day they were all gone. He was as excited as I had been 20 years before. They had to have come from miles and miles away for there to be that many crows.

Mother Nature is amazingly divine!

All I said was "Albert, why do rocks roll down hill?"

Dominant wife.

Cancer.

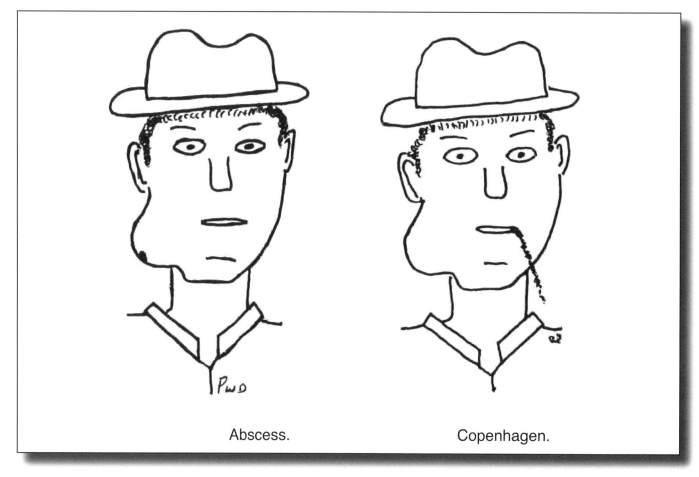

Abscess.

Copenhagen.

You and your old "Let's go north early and get on with the mating season" is getting a little old.

I don't think the Magnetic Mattress Pad is installed quite like that, Mabel.

Clyde, looks like we better reduce the canine genes on the next batch of Homo Sapiens.
We are getting way too many head injuries.

I say old chap, but have you got any Gray Poupon?

I have no idea where doc went. He just mumbled accounts receivable, emergencies, kicking cows, farmers complaining. He pulled the calf, took his chain off and…poof…he disappeared.

No! No Dear! I said I would spend $600 on a
De Laval Alpha Baby, not on a $600 purple Beanie Baby.

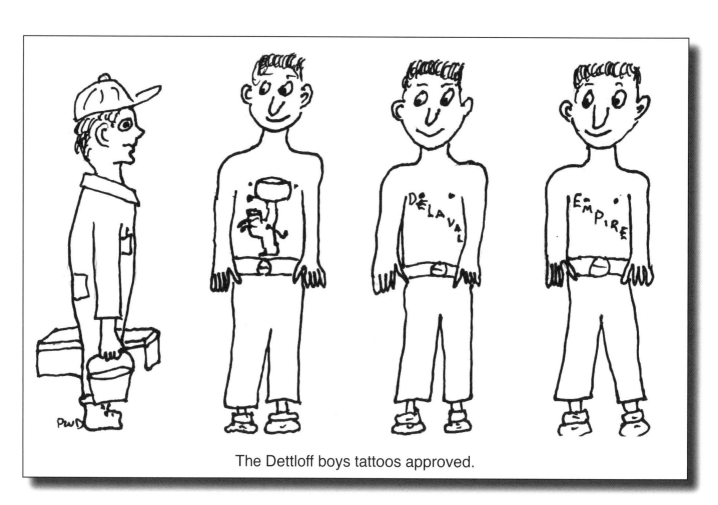

The Dettloff boys tattoos approved.

Cute little buggers, Clyde. Wonder what they eat.

The Horse

There once was a horse that could play a guitar,
So he headed to Nashville to become a star.

He was as good as Chet Atkins, and Roy Clark he
    was as fine;
This guitar strumming horse was a talented equine.

He would play the "Orange Blossom Special" just
    to warm up;
"Wipeout" was routine, he could play that at a
    gallop!

He got a job in a Nashville bar – the crowd they did
    cheer and shout,
Then one night he was discovered by a record
    company talent scout!

He signed the horse up, the label was M.C.A.;
He said, "Come in tomorrow, we'll tape you all
    day."

The first song released was a hit, a real snappy
    discord;
In less than a month it was #1 in the Magazine
    Billboard.

Then came the albums of beer drinkers, love songs,
    the ballad of a friend,
In less than a year this four footed strummer was a
    legend.

His wardrobe was flashy, all western, with sequins –
    Oh, what a sight!
He was invited as a guest on Johnny Carson's show
    called 'Tonight.'

Then while putting on a concert for thousands to see
    his fame,
The strumming, singing horse pulled up real lame.

Oh! What to do? There's much money out there to
    net.
His agent got on the phone and called an equine vet.

The exam revealed a pulled tendon and a hoof with
    a stone.
He was put on Butazolodin, Banamine and
    Cortisone.

Now, it's too bad our rich equine, his stall lined with
    rich rugs,
Before he can perform, he has to get loaded with
    drugs.

It's gained him nothing, his talent, money and fame,
As the vet says, that poor animal will be permanently
    lame.

So, home he went. Left the lights, glamour and
    charm;
Went back to the old homestead and tranquility of
    the farm!

I'm telling you, a properly balanced, biological farm takes the work out of farming.

How are your Van Gohs for grazing ability? Fine, but they can't hear worth a hoot.

You don't happen to know a good shamen any place?

I think it might be wise to quit using Bag Balm on that sore arm and see an M.D., Clyde.

Since this farm went organic, I'm sick of garlic tincture,
garlic for pneumonia, garlic for mastitis, garlic for this, garlic for that.
Just once, I wish they would try peppermint or malaleuca.

Prehistoric Man

Now, my family and I hunt Indian Artifacts along the
 creeks and streams.
I'm a lucky father, as I've got kids that lust for
 Clovis Points in their dreams.

We'll be hunting along in the spring, after a hard
 rain,
First item you'll find is a quartz triangular point,
 neat, but they are plain.

"Hey, Dad! I got a flint knife – it's well worked and
 grey;
I think I'll keep hunting, this could be my lucky day."

Next, my young daughter finds a Woodland pottery,
 a nice little chard;
She just walked past a nice chip, I think she's trying
 too hard.

"Let's keep walking and looking, kids. Oh! It's such
 a thrill
To know that on the next row might lay a Paleo
 drill."

We examine each chip closely, to see if they did
 flake.
Why, who knows, the next piece might be something
 archaic.

Ahead, the boys move, those guys are always a
 rushin';
They just found a Hammerstone, all battered by
 percussion.

A corner-notched arrowhead found next, a beautiful
 chert;
It was lying in the middle of the corn row, on top of
 the dirt.

My older daughter and mom, now they continually
 talk;
I just picked up a crude scraper that right past they
 did walk.

What kind of man made this nice point and sharp
 blade;
He fed and clothed his family; they were the tools of
 his trade.

These people were actually ingenious, so cold and
 alone;
They had a culture and life that was centered on
 stone.

So home we'll head, with some forgotten chips and
 artifacts of time long ago;
And catalog and label them, so our friends we can
 show.

34

Thunk…

Ugg Mehrings first Squash-em milking machine.

Clyde, we have got to do something about the environment. The Smiths are full of heavy metals, the O'Neils have dioxins in them. I'm so sick of eating overweight tourists. They just don't taste like our homegrown stuff.

This child obesity problem has sent my cholesterol up 40 points.

Been hunting grizzlies again, Tonto?

Any contact with other dogs lately?

I told you these aren't "Paleo," "Archaic," or "Woodland" type indians we're dealing with.

Try shooting to the left, or from the ground, Short Cloud.

Only Three

Science has as of yet not a clue,
As to what is in those tinctures of cobalt blue.

There are only three items in the universe this day;
Protons, neutrons and electrons spinning away.

The elemental chart with its orbits of energy,
Centrifugal and centripetal make up you and me.

When electrons move in and out of their orbital
    spheres,
Energy is lowered and raised over the years.

When medical men take tinctures with molecules
    abound,
And dilute past avagodro's number there's no
    molecules around.

This doesn't work; it's a goner they say,
It's only witchcraft in this double blind scientific day.

They don't understand the energy frequencies are
    transferred along,
When the original substance in many dilutions is
    gone.

We will discover the electromagnet template or
    energy pactet we've got,
Will play a huge role and stimulate the immune
    systems lot.

Science condemns, medicine scoffs and the FDA has
    the final call,
But a day will come when we will be smart enough
    to understand it all.

To me it is so logical that the basic three when we
    understand fully their action,
The proton, neutron and electron of the atom will be
    the main attraction.

So I will keep tincturing, stirring, combining atoms
    and letting nature synergistically help me,
As I have seen the results of tinctures and the
    energies of homeopathy.

He says if I get a good clean Sharples separator or a Rare Bird
in the next seven light years to let him know. He's interested.

Let's boogie outa here. This place is a mess. Athletes get millions. Two percent produce food and can't control it. One third either work for, or are supported by, their government. Scratch this one off the list.

Holy Moly! We've hit the Mother Lode.
Let's dispense with the formal intros and get to know these folks.

That sure would have been a nice one next year.

Well, mine will be good eating.

I just hate it when Doc gets a deer.

Reed Canary Grass and the Deer

It was Saturday night in the 70's. I had worked on Saturday as that was always a big day practice wise and income wise, and there was help at home.

I was on call Saturday night and Sunday as it was my weekend to work. My in-laws were invited out to our house for supper and cards, as I married into a family of fun loving card players. I had gotten home about 5:30 p.m. from my last call, washed up and ate a wonderful meal my wife had prepared for us, the kids and the grandparents.

We just sat down to play cards, and the phone rang. I had a milk fever east of us on a farm close to my last call. So off my father-in-law and I went. We got about half way there, going through some wet, brushy area and here lays a big doe deer, dead on the little narrow shoulder. I stop to check this out, as she was not there 90 minutes ago.

She's still warm and appears to have gotten hit in the head, breaking her neck. This means she has got about 70 pounds of nice deer meat on her 140 pound body. At that time it was not an approved procedure to harvest highway deer, as they belonged to the DNR. They have since changed that so if you contact the authorities and wait long enough, you can then fill out a form, give them $5 and you have just bought yourself a DNR dead deer.

Not being a person with patience, and knowing it was not proper to butcher these DNR animals, posed a fun challenge to my father-in-law and me. We pulled her down into the ditch, I got out some dandy post mortem knives from my veterinary truck and Pops and I had her gutted out in about six minutes. Only one old grey haired lady went by, so we were pretty safe.

I said "Let's leave her here in the ditch, hidden behind this tall reed canary grass, and pick her up on the way home."

We washed the blood off, put the knives away, and discussed butchering her that night after we returned home. We sped away to the milk fever. The cow responded really well and we headed back home, licking our venison chops all the way.

We pull up to the ditch, my father-in-law jumps out of the truck and headed down into the ditch and starts swearing. You see, my father-in-law was a wonderful man, but he was a professional swearer in his everyday language. I think nothing of it until I get down into the ditch and there is no deer!

Gone! Our deer is gone!

My father-in-law looks at me and asks "Didn't we gut out that ****** deer just an hour ago?"

I said, "Look at the yard lights (which we could see about 5 of them). Someone is laughing at us right now. They hit the deer and went home to change clothes and get a knife, came back and found the deer gutted out. Let's go play cards."

We did.

2 A.M. Milk Fever

Being a large animal veterinarian, you get used to the phone ringing anytime day or night. Usually you don't get your first call in the morning before 4:30-5:00 a.m. However, there are exceptions to everything.

Phone rings at 2 a.m. "I've got a good old cow coming down with milk fever. Please hurry."

Quite routine for this particular farm as the dairy cows are milked and managed by the wife. She happens to be on top of things, so no waiting until 5:30 or 6 a.m. to treat this milk fever. It's 2 a.m. – come and treat my cow NOW!

I've never had trouble falling asleep while driving on late night calls. The problem was always going back to sleep when I returned home. So, I head out. It's about an 18 mile drive. I get there and sure enough, she's coming down with a classic case of milk fever after freshening. It's true that the earlier you treat them the quicker they respond and the less complication one will have.

Now, at that time I was practicing out of a Dodge mini-van. Dodge came out with the first mini-van in 1984 so this was 1986-1987, somewhere in there. They had a back door that flipped up and I had a special made veterinary box setting there that worked good. I usually left the back door flipped up at the farm and shut it when I was done and ready to leave. (Unless it was winter, then everything was closed up all the time!) This I had done as per usual on this particular night. I got paid, crawled into my mini-van and headed home.

Now, these night calls were always sort of a little reflection time for me. I never turned on the radio and I always checked to see what phase the moon was in and located the Big Dipper and the North Star. I would fold a towel up and put it against the window and drive home with the road vibrations coming through the towel in sort of a relaxed quiet contemplating state. I was cruising home so sedate, when all of a sudden the fastest most scared cat comes from the back of my van, hits the windshield on the right side, does a 90 degree left, zips across my dash spewing everything on the dash all over my van. He hits the left front corner, does another left and flys over my left shoulder returning to some hidden area in the back of my mini-van. This all happens in less than 10 seconds. My calm tranquil moment is over.

I get an adrenaline rush (Cortisol rush) and come to a quick stop (slammed on my brakes) on this country road. I go to the back door, open it wide and step back.

"Here kitty, kitty, kitty. Come out kitty, kitty."

I know it will take some time so I do what any normal man would do, I take my time emptying my bladder. I'm just done and I see this orange streak exit the back of my mini-van. I'll tell you that cat was moving. Through the ditch and into the woods. Good-bye kitty!

I never did say a word about it to my farm client, but I often wonder if that cat ever made it home. I did, and no, I could not go back to sleep.

I said catgut, you dimwit. Can't you hear?

Doc really treats his milk fevers slow.

Yes, Ma'am, we will certainly call you if we find your Fifi.

My First Dog Spay

I bought a big two-story, four bedroom house in 1967 with an attached garage and a full finished basement.

The basement I turned into a little small animal operating area. Back then, country vets were expected to do all facets of veterinary medicine.

A jovial, short, pudgy, older lady had this little circle dog to spay. It was a Pekinese that barked and ran in circles. She bought it in Winona, Minnesota. I performed the spay. Upon picking it up and paying me – which by the way was $15 (castrations were $10 in 1967), I said the stitches should come out in three weeks. Her husband, short and jovial also, chimed in saying he had two Holstein male calves that needed to be castrated so could I come out in three weeks and take care of both the bulls and the stitches. I said that would be fine and we set up an appointment.

On the appropriate day, I drove in and saw the calves in the pen and took care of them first. I had a pickup with a veterinary unit on the back that had a back door that flopped down, making a handy table. We set the circle dog on it and I snipped out the stitches. I put the dog down to watch it wind up to do more circling. The farmer and I started to visit – getting to know each other. He asked me some questions as to where I grew up and how it was being a farm boy in Minnesota. We had a nice friendly exchange and I said I'd better get to work and it was nice to meet you, thanks for the business, etc.

I put my Dodge in gear (I had a three speed) and started to go. Little did I know that that little circle Pekinese had never been out of the house and had developed no concept of what a tire could do to you. SQUASH – I ran right over the dog's mid-section, bulging out it's eyes, popping out it's rectum. An instant death. I immediately knew what I had done. I jumped out and joined the farmer as he was looking at his dog. He looks at me and said, "Well, your stitches held." (I love positive people!)

We both went to the house to explain what happened to his wife, leaving the dead, non-circling dog outside.

His wife went to Winona, Minnesota and bought a new dog which I spayed for free and threw in a vaccination for good measure.

Always, always know where circle dogs are when you leave a farmyard.

Just my luck. About the time I could use a hand, Clyde's going to check out his cornfield.

Pull, Erma, pull!

Yep, old Doc Snead always took his shirt off on those tough ones.

Wait until he's in the barn. I'll get the front two, you get the rear ones.

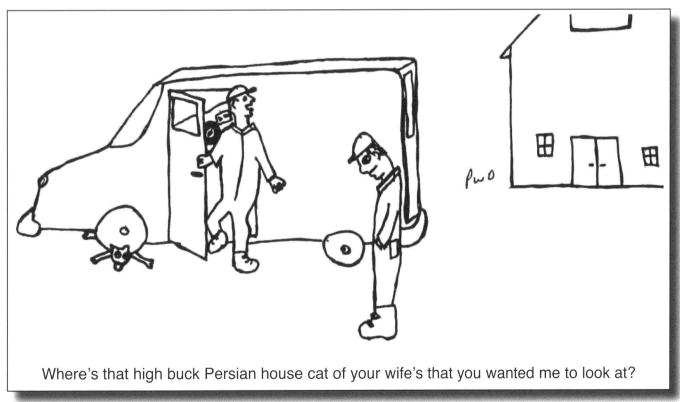

Where's that high buck Persian house cat of your wife's that you wanted me to look at?

Obstetrics

As a veterinarian, delivering calves was one of the most enjoyable challenges.

You had to keep in mind that there were only two times you could put a lot of pressure on pulling a calf. One is when the calf's head is laying on its front feet and one is when the back feet are coming backwards with the tail placed above it. You always wanted to ascertain that the two legs belonged to the same calf. You did this by going up one leg to the tail and back down. A caution is to make sure the umbilical vein and cord from the calf's navel has not gotten caught, by usually the left leg, and goes to the placenta for sustenance. When you start to pull, you shut off the lifeline of the calf and you have about three minutes before the calf will drown in amniotic fluid. Everything you're manipulating previous to those two positions is to get the calf in one of said presentations.

There are head flexion to the right and left; there are feet back, one or both. Twins are a challenge; it's usually best to pull the backward twin first. If you don't know where you're headed, and can't figure out what you're doing, don't mess around. Get help from someone who is trained. The longer you mess around, the further south she goes health-wise.

I had two brothers that didn't have a real high IQ that were pretty tight. When I went there it was usually too late for me to be of any help. One night about 1 a.m. my phone rang. One of the brothers was calling in desperation as they had been working on this poor cow since 9 p.m. and couldn't deliver this poor calf. I said leave her alone, go in the house, and take a break. They didn't. When I walked into the barn all the cows were out. This poor cow was down. One

brother was on his belly on the walkway with both hands in the cow. He was red-faced. The other one was on his knees, drenched in uterine fluid. He had been pushing his hair back, so it stood on end. As I walked towards them the image of Jim Carey and his partner in Dumb and Dumber crossed my mind. I got some warm water and offered it to the cow and she drank it, as she had been through four hours of miserable labor. I rolled her onto her side instead of flat out and slapped her back and yelled at her and she got up. I reached in and felt a uterine torsion. I really lucked out, as the fluid was gone. The uterus was contracting down and I gave a clockwise flip with my arm around the head. I put chains on and had the dead calf out in about six minutes. They had no idea what a torsion was. Most of them are not that easy.

Torsions are when the entire calf and uterus flip 180°, most always counterclockwise (except in the southern hemisphere, as that's how our earth spins. Same way with whirlpools with water.) Torsions were rare the first 10-12 years of practice. Now my last few years they have gotten real common, especially in the high production herds. I feel it's a factor of demineralization from the soil up. Good high forage herds have very few torsions.

When pulling a calf always double half hitch the chain or rope so you don't break a calf's front leg. Work with the cow. Take your time. Lubricate well and watch her labor. When she pushes you pull. Take your time. Too many times young vets panic and are way too fast. I always gave the cow warm water after calving and put a homeopathy called Pulsitilla in the water to aid the mother in passing the placenta. Always, always let

the mother lick the calf off. This is a precious time for the spirits of life. It helps the cow and the licking stimulates the calf.

I've delivered many, many twins and three sets of live triplets, two sets were all females. Only once did I deliver four calves from an average sized Holstein heifer. They were premature, about 25-30 pounds, all dead. Her uterus was huge. She had a mountain of cleaning and was so weak she crashed and died in about two days. Four calves is extremely rare. This was before hormones also, so it was natural. That was about 1968 or 1969 that I delivered the four.

About six weeks into practice, I delivered a Shistosomus Reflexus calf. We had one brought in to vet school when I was a student. What is a Shistosomus Reflexus? Well, it is a calf that has all its organs, but the thorax and abdomen never close, so all the organs are dangling free. The heart and lungs in the chest, and the stomach, intestines, all the bowels, bladder and abdominal organs are there in plain sight. There is usually a spinal defect about where the thorax and abdomen meet at sometimes close to a 90° angle. Most of the time you can pull them. I delivered five of them all early on in practice, mostly the first 15 years. My last 20 years not a one. Most young vets have never seen one. I'm thinking it was some genetics causing it and we lost those genes. They were also all in Holsteins. There are quite a few facial defects. Especially in Holsteins. Also ankylosed bent stiff legs, front and back. I've delivered three two-headed calves, usually a combination of three eyes, three ears. Grotesque looking heads.

Only once did I deliver a perfect two-headed bull calf with four nicely placed ears and four eyes. I diagnosed it before I delivered it. It was on an Amish farm. I was pretty good friends with the Amish man and I said I want to buy this calf after I deliver it. He asked why and I said you will see. I tilted it and got one nose through the pelvis and pulled it sideways and got the other nose to follow and pulled it out. It was mostly black and was perfect in all other respects, except it was dead. He gave it to me and I have it to this day, mounted. People ask if it was born alive and I say 'Oh, yes, but such a tragedy followed.' They ask what I mean. I tell them the guy on the right thought the guy on the left was eating and the calf on the left thought that the guy on the right was eating and they both starved to death.

I delivered a heifer calf with a fifth complete leg on top between the shoulders. It had no scapula so it just dangled there. When the calf was about five months old, the farmer had me remove it, which was not real difficult.

If you cannot deliver the calf, a C-section is always an out if you have a live calf. If you do a C-section with a dead calf it's hard to save the cow. I did very few C-sections in my last 10 years of practice, as experience is a valuable teacher.

Those poor buzzards will never get a fertile egg as long as they keep texting their mating call.

Vet Talk

It was 10:30 at night, 25 below, dark in back, and I finally had her uterus about where I wanted it and then she rolls over on me and pushed.

Supper

It's six p.m., I'm rolling in late, weary, hungry for roast;
Ready to have a big meal, lay back and coast.

The oldest daughter home from college, pursuing her degree;
The tall one's her boyfriend, shooting baskets and throws that are free.

The second daughter needs a taxi, the trips never end;
Tonight she brought home a cute volleyball girl friend.

Oldest boy's limping around, that kid sure can moan;
Got stepped on in football, hope he didn't chip a bone.

Second son just did his homework, pulls up a supper chair;
Better knock off early tomorrow, he's got Junior High football at Blair.

Youngest daughter, she's a high energy case;
She's sampled the roast, as there's catsup on her face.

The third boy and youngest, he's about in pre-school;
For some reason he's in the bathroom, standing on top of the stool.

My wife yells 'Supper!', time for the utensils to yield;
At certain points in time, the kitchen's a battlefield.

'Pass the bread,' 'I need milk,' 'Send me the peas,'
'Hold it,' I yell, 'Can't anyone say Please?'

'You got my fork,' 'We're short one glass,' 'Please pass the salt,'
Elbows are flying, plates are whizzing, you'll starve if they halt.

'Now, just a minute, boys, you took too much meat.
We've two extra tonight, just let me have one piece to eat.'

The two guests are stunned; they're simply amazed.
The girl friend is bleeding, her arm, it got grazed.

Oh, was that good, we all lean back. Mom had another winner.
After it's over, we all thank her for the good dinner.

All the friends love a big family, as here's where they all congregate.
I think next time around we'll put up a sign, 'Motel 8.'

But, so you know what, and it sure isn't fair?
I always feel empty and incomplete when there's a vacant chair.

We do have a Mexican entree on the menu tonight, Carib.

50

Yep, Doc! Clyde crossed his cows with Great Dane dogs and, with his new fire hydrant milking bench, milking is a breeze.

Smart Pig

There once was a pig that had two heads;
He had two pillows, he had two beds.

He had four ears, two in a clump,
Long skinny body, little round rump.

The other pigs would eat all of the time;
He liked his slop with a little salt and lime.

His buddies, away to market would go;
He was always too light – when to sell, I don't know.

The other pigs while eating would say, "You try it."
"No thanks, fellas, because I'm on a diet.

You guys slow down, you will end up as pork
In some fancy restaurant in Boston or New York.

I'm staying away from that butchering train.
As you see, I'm a pig with more than one brain."

One day they loaded him on a truck headed to Hormel.
He only squealed once, but "Oink! Oink!" he did yell.

51

I'm told everything tastes better bacon wrapped.

You tell me you don't want sweet and sour or cured ham,
so what's a plastic surgeon got left for choices?

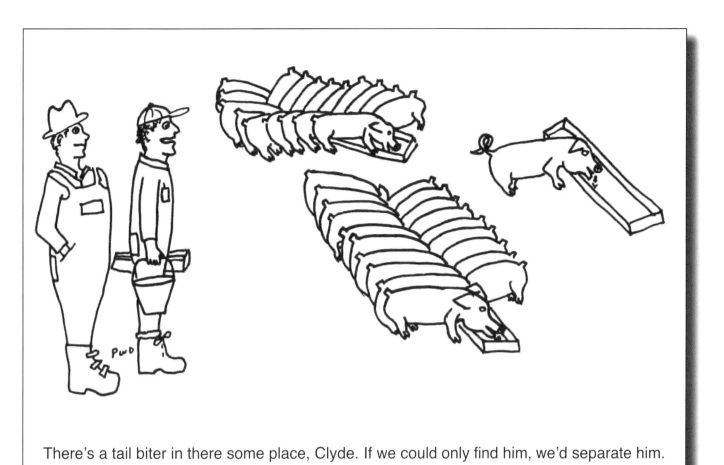

There's a tail biter in there some place, Clyde. If we could only find him, we'd separate him.

Only you would pick a wooded forty that's scheduled
for clear cutting to set up housekeeping in.

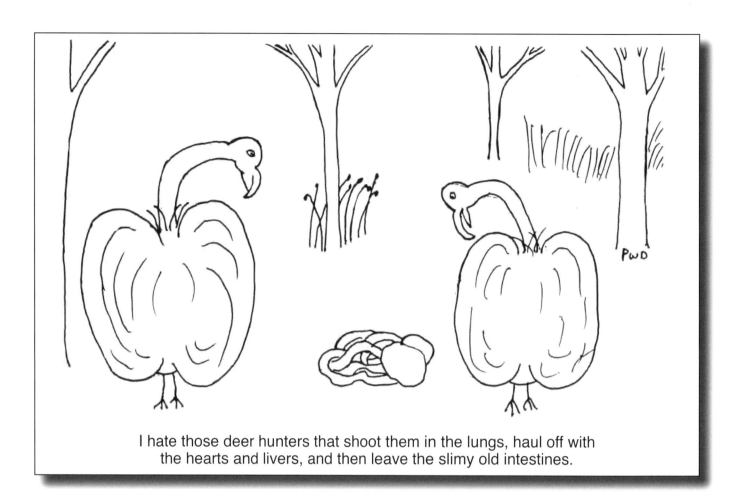

I hate those deer hunters that shoot them in the lungs, haul off with the hearts and livers, and then leave the slimy old intestines.

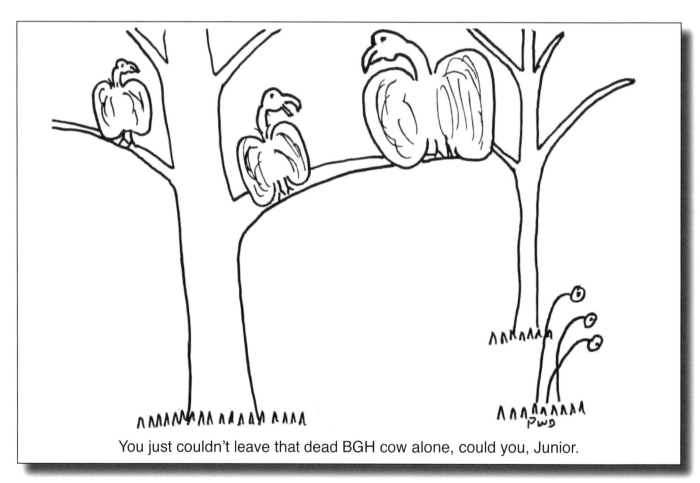

You just couldn't leave that dead BGH cow alone, could you, Junior.

I said I want you to bring herbs and whey to use on my mastitis problem, not curds and whey.

Some vet he is! Doesn't know how to pregnancy check a marsupial.

Wants And Needs

I was raised in the forties, a Minnesota dirt farm boy.
We had clothes, food, love, and for Christmas got one toy.

People worked to fulfill their necessities and needs.
Today we run to the malls to break the tension and satisfy our greeds.

A demarcation we've lost, and to which we pay no heed,
Is the difference between a want and a need.

I'm no different than the rest, I've got lots of wants on the shelf.
I don't really need any of those items that contribute to my wealth.

What's it doing to our offspring? All the material things for granted,
Different than the conservative values that my parents in me planted.

Our society has changed so very much,
I just hope our values stay in touch.

It's a circle, no one keeps things forever, it won't fail.
After the funeral the estate's distributed by an auction sale.

Wants and needs should be more defined in school,
So that we could grab those values more like a tool.

A need is a necessity, like the food of life.
A want is a satisfied state, like a happy, loving wife.

Needs you can purchase to keep on living.
Wants are something that should require giving.

When my offspring say, "I really need that, Dad,"
I quickly explain, "That is a want, my lad."

I feel very fortunate to live in a time when I can easily fulfill my wants,
And have time and money for pleasure and can frequent fun haunts.

I said one cupcake for desert, not three, Viper.

Wisdom

Wisdom comes orally from some quiet, crusty old fella that's sort of nifty.
You don't possess much of it unless you are at least fifty.

Wisdom is failures, mistakes, experience and keen observation,
Mixed with common sense, logic and knowing what's happening in the nation.

It's knowing you're right, and against odds and opinions you do it.
To have the persistence and drive to see yourself through it.

Wisdom is not to be dispensed too often to every wandering soul,
But to be clearly delivered to a listening ear when someone's in a hole.

A simple comment with no yelling or fist shakes
Can accomplish much. Sometimes that's all that it takes.

Young fellas compensate with enthusiasm, drive and dreams.
Old fellas with wisdom conservatively see the holes in their schemes.

But everyone survives life's different eras of prime.
I say wisdom takes fifty years of life's time.

Looks like we better thin out a few cavemen before they eat all our bacon.

Rightfully, it does belong to Mick. He earned it.

You forgot the D-Con again, Harold.

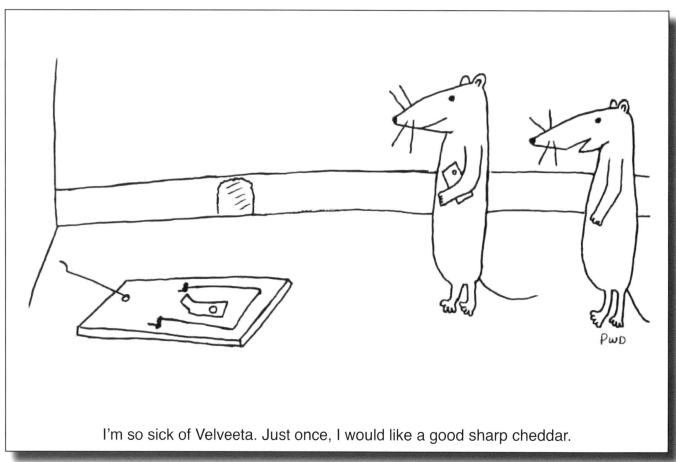

I'm so sick of Velveeta. Just once, I would like a good sharp cheddar.

Nursing Home Woes

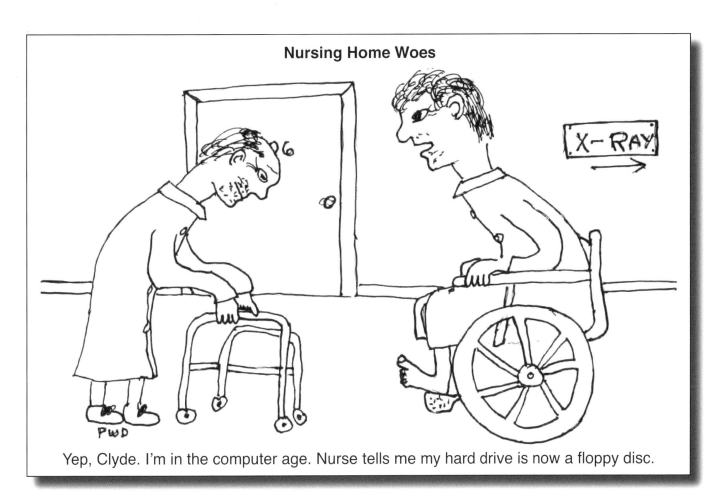

Yep, Clyde. I'm in the computer age. Nurse tells me my hard drive is now a floppy disc.

Nursing Home Woes

Nursing Home Woes

If you wouldn't have collected those heavy cast iron cream separators, you could stand up straight like a normal old collector.

Nursing Home Woes

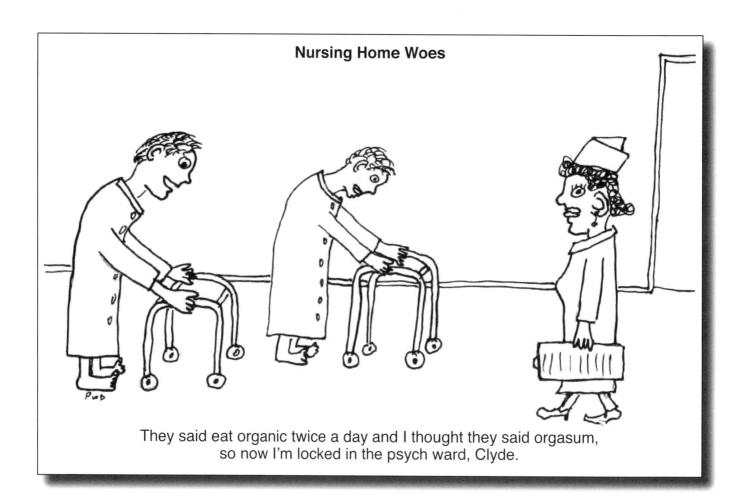

They said eat organic twice a day and I thought they said orgasum,
so now I'm locked in the psych ward, Clyde.

Nursing Home Woes

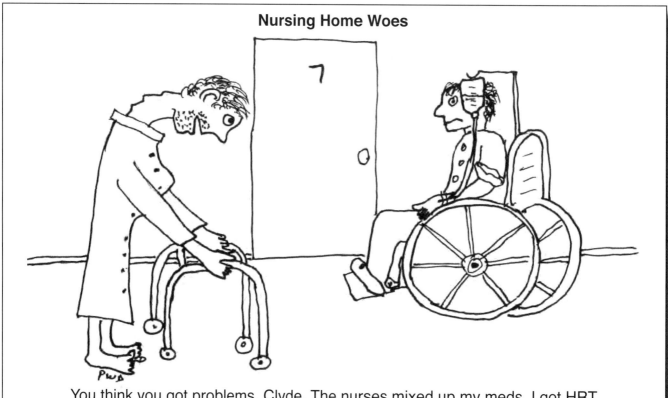

You think you got problems, Clyde. The nurses mixed up my meds. I got HRT
instead of my cholesterol and blood pressure medicine. And now I grew breasts
full of cholesterol and I feel bloated and bitchy every month for a week.

Nursing Home Woes

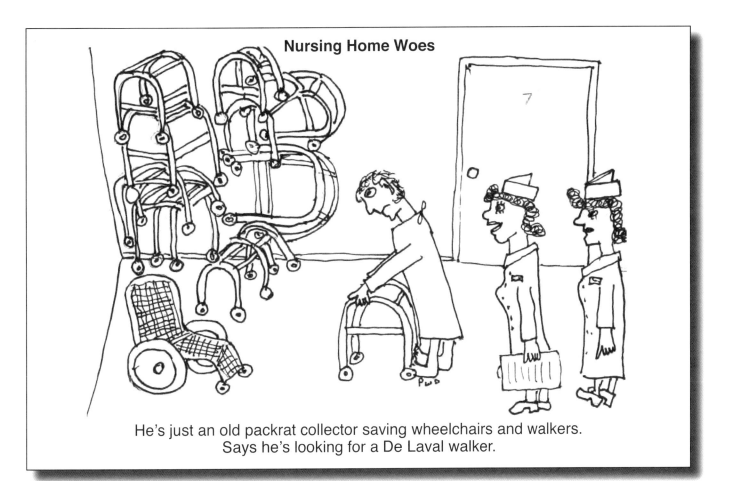

He's just an old packrat collector saving wheelchairs and walkers.
Says he's looking for a De Laval walker.

I told him that we should avoid this Clovis area.

Collecting

I collect old iron, separators, and once in a while an
   antique.
The other day, I found an old tractor down by a creek.

Most farmers won't sell, or want way too much;
Now I'll be looking for an old tractor clutch.

I'll buy that cream separator, leaning against the bin;
It was a rare old De Laval – a No. 10.

'No way,' he said. 'That belonged to my dad.'
'No problem.' I said, trying not to get mad.

For twenty-two years I was his vet, I'd been around.
I know it was longer than that, since his dad's in the
   ground.

'I'll sell you that old thresher, made by J.I. Case.
You've always wanted it over at your place.'

'That sounds good, I'll find a spot for it to stand.'
'Pick it up next week, I want six grand.'

It's tough to buy local, because I'm a vet.
Everybody wants double, and then some, you bet.

So before the auction, I call the clerk and put in my
   bid.
Then they all think it was bought by the neighbor's
   kid.

Pack Rat

A minor problem exists at our happy abode.
That's when things get thrown out, I check the load.

You see my wife is a tosser, we don't need this, we
   don't need that.
Unfortunately she is married to a conservative pack
   rat.

Routine household garbage usually ends up in the
   landfill.
But, when my wife cleans out a closet that's surely a
   thrill.

Here's twelve inches of gold colored chain of a worn
   out lamp.
That we'll save. Hang it in the garage next to a clamp.

This good old toy, only the wheels off, to toss it is a
   shame.
I hope I don't get discarded when I come up lame.

We'll put that in the box with the other broken toys
   on the shelf,
When I retire, for the grandkids I'll fix them myself.

Now look at this waste, here's a fine old corduroy
   coat of my dad's.
Why just the other day I saw the same thing on a
   couple young lads.

A broken scissors, well what do you know?
Why, I've got another half saved someplace, now
   where did it go?

When the basement gets cleaned, well that's a real
   hey day.
I just can't believe what my wife throws away.

She's getting clever and sneaky, using a grocery
   sack,
She'll try and get rid of it behind my back.

But I watch pretty close, I know most of her tricks.
Why once she hid good stuff in a bag of leaves and
   sticks.

My glory is when she says, "Dear, you wouldn't
   happen to have…"
I head to my treasures with a grin and a laugh.

And when I return with the needed item and it fits to
   a tee,
I say, "You know, you nearly threw that away on
   me."

Now I'm a little older and will probably leave her
   first,
So I regret, but my goodies by auction will be
   dispersed.

I just hope that they will be bought, I don't care if
   skinny or fat,
Just as long as they're stored, saved and enjoyed by
   another pack rat.

I don't like coming here, so I make this guy pay through the nose.

Can't lose her, Doc, best cow in the barn.

Chute Jobs

One task in the veterinary profession that doesn't
    require any great mental shock,
Is what's called a chute job for beef cows, beef
    calves and young stock.

The horns get removed, be sure to pull out the bleeder,
The jewels get dropped, so they won't be a feedlot
    breeder.

Grubs, lice and worms are all attacked with pour on
    injectable wormer or Ivomec.
No pity on those little vermin, they take it in the neck.

Pregnancy checking to see if there's a calf, with a
    sleeve.
Vacinations of Lepto and the viruses so the fetus
    doesn't leave.

Implants put in the ear to get more efficient gains
    per day.
They are now all ready for a feedlot in central Ioway.

Some jobs take a couple of hours or more to run
    them all through.
So I converse with the workers and owner about
    what's new.

We cover government programs, corn base, set aside
    and C.R.P.
I get the philosophy, world views and sometimes an
    opinion on me.

How did we do deerhunting? They got a dandy ten
    point rack.
Do I think the coach for the Packers, next year, will
    be back?

I ask about the neighbor's new house, must have cost
    a hundred grand.
They grumble and say he hasn't paid last summer's
    hay rent on their land.

Should be a good year for your boy in wrestling, last
    year he went to State.
About that time one wild Angus smashes a steel
    Kentucky gate.

I see you got yourself a new four wheel drive John
    Deere.
Yes sir, milk price wasn't too bad, we had ourselves a
    pretty good year.

How about you, Doc, any winter vacation coming
    up, some neat place to go?
Yep, me and the wife, in February are flying to
    Cancun, Mexico.

I guess you can call it in our vet circles a strong
    rapport.
One thing we have is a good client relationship, that
    is for sure.

The M.D.s only see their clients in their own medical
    scene.
The vet sees his clients in their own habitat, shabby
    or serene.

So, in the fall I keep doing those chute jobs until
    evermore,
I'll keep right on talking and visiting and building
    my rapport.

Acute Pancake Head

I had an unusual client that called me once a year for a chute job. This was an older brother/sister team. They had a lot of land and other real estate, and in their late autumn when they were too old to milk, turned their farm into a classic wild beef cow (by neglect) operation.

He was a big, jovial, neat sense of humor fella that called everybody "Governor" (at least he did my partners and me). She was a quiet, older spinster whose whole life

was taking care of her brother and her cats. Her cats were definitely No. 1 and her brother was No.2.

So he calls and says he has 27 head of 6-12 month old calves, mostly bulls that need to be castrated and dehorned. These 27 head have just been weaned, have never been touched by a human hand and haven't seen any humans other than their caretakers.

I arrive at the farm and he meets me and tells me to back up to the usual door, where I was last year,

next to the silo and old milk house. As I'm backing up, I see quite a few half grown cats scamper hither and yon. He hollars "Whoa," I stop and I made a perfect back up right square to the door. While I'm unhooking the chute from the door, Sis is scurrying about, moving the dishes she feeds her cats in and talking to her little cat colony. Being very timid and reclusive, when I get out to set the chute she disappears into the house.

The way a cattle chute works, is

both wheels are on a swivel like apparatus, where by you stand on this lever and unhook first one wheel and set the chute on the ground on one side, then do the other side so it sits on the ground so it can't slide forward when cattle are in it. Upon putting the second side down and being a heavy thing, it usually goes down with a lot of weight. Ka-thunk. The brother was standing in the barn by the back of the chute when I let the second wheel down. From under the chute we heard this loud cat scream that we both knew was the scream of death. There was a little scratching on the metal chute bottom, then complete silence.

I quickly jumped on the lifting lever and raised it as quick as I could. He got down on his hands and knees, took one look under the chute and said, "Put her back down, Governor." and went to fetch the first calf.

We worked for about 90 minutes on his cattle and got done like usual. I then pick up one side of the chute, then the other side. He snatches that pancaked cat out from under the chute and disappears quick as a flash into the barn.

He always paid with a check so I pull ahead, get my billing machine out, and write up the bill. Sis comes out of the house with some more cat food to feed her cats. We're about 20 yards away, standing by my pickup and we both notice she is calling her cats and looking here and there. Finally, as I'm packed up and about to leave, she asks her brother if he's seen Fluffy, as she's not around. Her brother said, "Oh, she probably took off when the Governor here drove in with his chute. Wasn't she that wild little one?"

As I leave he says, "See you next year, Governor," and he heads into the barn to take care of his pancaked cat.

Hold that thought, dear, while I go get a stool.

Is yours a cross your heart or a Playtex?

Doc, my vet bills are so big I'll never be able to pay you.
Why don't you just marry my daughter, Isabel?

Well, they finally got a kid with his head screwed on straight.

No, castrations are not expensive. Which pen is it in?

What Happened That One Year?

When I started practice in 1967, there was an Old Order Amish settlement east of me about 15 miles. I started doing some veterinary work on a few farms. It didn't take me long to like these folks. They were very logical thinking, very close to Mother Earth, the moon and nature. Plus, they paid their bills – genuine, honest people. We could all learn a few things from them, if we'd listen.

One evening, the phone rang. It was an Amish fellow, calling from a neighbor's phone, telling me he had a good old Jersey that had just freshened and was down with milk fever. It was winter time so the cow was in the barn. She was in a fairly big pen that was filled with deep yellow straw. Her head was around to her side, she was down getting fairly sleepy from lack of calcium in the blood stream. She was a text book milk fever. They go down because calcium is needed for muscles to contract. She had used up a lot of calcium, as the uterus is one big, smooth muscle that worked hard contracting down during the birthing process. Her udder also filled up with calcium, rich colostrum and milk. Blood calcium can only come from the food eaten that day or the skeletal system. The bone calcium is controlled by the parathyroid gland. If that is a little slow and can't keep up with the calcium levels needed in the blood, they go down physically.

I secured her head by tying it to her leg and inserted the IV needle in her jugular vein and started running in 500cc's of sterile calcium and glucose. This takes awhile so this is a good time to visit.

I looked over and my Amish friend was sort of a little guy with a long beard. Beside him, in the lantern light reflecting off the yellow straw, stood five young boys. They were lined up stair step fashion from oldest to youngest next to him. Having a veterinarian come is a big event for an Amish family and his boys were savoring every minute of it. I asked them their names and received five biblical names. I asked the Dad, "How many children do you have total?"

He answered, "Eight total."

I had noticed the five boys were probably only 5 years and 15 minutes from youngest to oldest.

"Well, how long have you been married?" I asked.

"Nine years," he replied.

Without even thinking, I blurted out, kind of jokingly, "What happened that one year?" He looked at me very seriously and said, "You know, I don't know what went wrong."

So, Roger, how long have you had your dog on Purina?

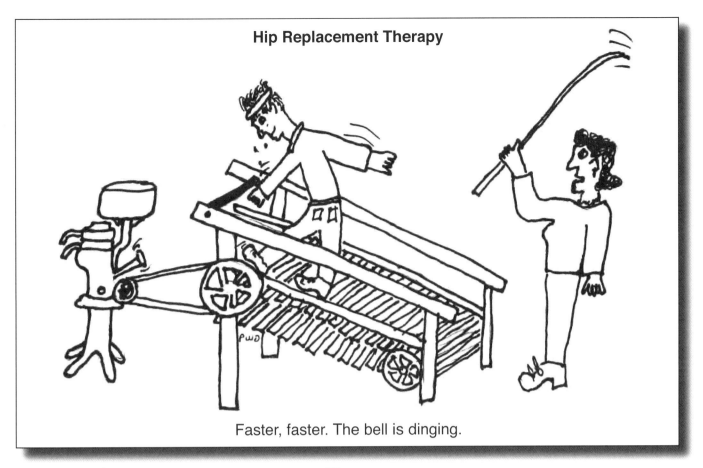

Hip Replacement Therapy

Faster, faster. The bell is dinging.

Dogs

Every farm has a dog, some have more, up to three;
I've made a little study of them, I'll tell you what I
    did see.

I see lots of little Healers – some Australian, some
    Blue;
They are all timid and sneaky, and some want to bite
    you.

Lots of big old Collies – they are quiet and shy and
    all look like Lassie;
They tend to chase cars, but I've never seen one
    sassy.

The little black Border Collie – they are friendly and
    serene;
For rounding up cattle, they have got to be supreme.

A few St. Bernards – they are loppy and lazy, you
    bet;
For cattle they are no good, they are strictly a pet.

The Labs and Retrievers bark and run – they're
    regular jumpers;
First thing they do is spray both of my bumpers.

Beagles and Bassetts – they are in the same group;
They hardly move from their straw bed, their eyes all
    a droop.

The biggest group are mongrels, from parents
    unknown;
They'll be laying by the milk house, a chewing a
    bone.

Mutts are like mongrels, but they have a lot more
    hair;
They usually are sitting by a stove, next to a rocking
    chair.

The German Shepherd has a menacing bark, his hair
    standing on end;
But with gentle persuasion, they are usually a good
    friend.

The names that are chosen are unique, here's a few:
Ralph, Zeke, Bowser, Lady are fitting, but Spot is not
    new.

Most dogs have a distinct personality, it's usually
    quite nice;
In over twenty-five years I've been bitten just twice.

69

Mabel, cancel Snoopy's vet appointment. He just got his anal glands expressed.

And don't you talk sheepish to me.

Vet's Wife

Back when she was young, and pretty as could be;
After some courting, she decided to marry me.

Her pappy said, "He's a catch, a large animal Vet."
I know to this day, she questions that wisdom yet.

"Yes, Honey, I'll pick up Suzie. Yes, sir, at five
    fifteen;"
About that time you're twenty miles away, with your
    calf pulling machine.

Johnny's got a game at eight, for the number one
    spot;
You're doing a C-section on a heifer from a feed lot.

The in-laws are coming for Sunday dinner, a
    seasoned roast rump;
You get an 11:30 emergency, from a one-call-a-year
    old grump!

"What! Your vehicle's shot – why it's not three years
    old;
Somebody's got to get to the Doctor with Suzie's
    cold."

"No, he's not here right now, he's out on a call."
"You think you had a rough day, these kids are
    driving me up the wall."

You finally go to Church, after an Emphysomatous,
    and sit in your own pew;
The old widows up front look around, and think you
    must be somebody new.

"Sorry, Dear, I can't talk to the salesman at three;
But order me some Dexasone, Calcium and
    Histavet-P."

"The checkbook is empty and taxes are due."
"No, Honey, I didn't spend any, I'm working 'till I'm
    blue."

She's my secretary, bookkeeper, mother and queen;
She bandages up kids, cooks, and keeps my coveralls
    clean.

In spite of the practice, we love each other a lot;
And she's just as beautiful, as when we tied the knot.

Every time we go to the Smith's, you end up putting your stupid foot in your mouth.

Farm Wives

A necessity on a dairy farm to have a good life,
Is the needs and benefits of a farm wife.

I've seen dairy operations simply come alive,
With spark acceleration and female drive.

She's up early, helps milk and get the cows fed,
Then whips up eggs, sausage and toasted bread.

The money is out for the kids' school lunch,
Then out to the barn to feed the young stock bunch.

When the barn's all quiet, the cattle with contented
    looks,
She can sit down and work on the record books.

She can drive a farm tractor and unload hay –
That's her aerobics and exercise for the day.

The bulk tank's spotless, the pipelines shiny and clean,
She hopes the milk inspector won't be picky and mean.

The kids get picked up after practice at school,
She stops at the farm store to pick up a tool.

On she works in snow, rain and mist.
Nothing's too tough to go on her list.

She's the farm's accountant and banker, and
    everything does balance.
Involved in civic duties, she gladly lends her talents.

Farm wives may have rough hands from the chores
    and washing duties,
But if you measure their hearts, then farm wives are
    beauties.

Frugality

I consider myself thrifty with my coin, as I labor,
    I don't spend much;
Others call me frugal, some call me tight, some call
    me Dutch.

But, this young generation of wastemakers, for me is
    getting out of touch;
They throw away all containers, and use three
    napkins for one little blotch.

Ever go to McDonald's and ask for just one more
    catsup pak?
They give you three handfuls in a company
    embossed sack.

There you sit with six French Fries, drinking catsup
    – what a taste!
Because you asked for one, and you're too
    conservative to waste.

My kids behind my back snicker, complain and let
    out hollers;
When I keep telling them, "You know, pennies make
    dollars."

I save plastic containers with nice tight fitting lids.
When OPEC quits fighting and oil goes up, I'll give
    them to my kids.

We compost the decomposable, so the soil will be
    rich and thick;
But how do you dispose of those mounds of plastic?

We burn and we bury, and old refrigs we pay to haul
    away;
Because we are all being policed daily by the E.P.A.

I'll get home tonight with my five gallon garbage
    pail full to the rim;
I've got sleeves, pipettes, empty bottles and one
    gallon can of tin.

So I'll keep saving and storing, because it's my
    upbringing I thank;
I just sold my aluminum cans and the money I will
    bank.

She's a little wild around vets, Doc.

Doc's never been very handy with the lariat.

The Way It Is

I worked for a bachelor named Harry McGruff;
His place was a mess, the yard looked real tough.

I didn't like going there, because of the mess;
My diagnosis was usually an educated guess.

They all got a magnet, that was for sure,
'Cause anything that did eat, passed nails in the
*   manure!*

One day I said, "Harry, clean up this place."
He looked at me, astonished in the face.

"What!" he said, "Have this look like a park,
And tie up my dog so he can't even bark?

I'll be so busy, I'll have no time to relax.
Then the County and State will raise up my tax.

I'm saving nickels and dimes, you see,
I'm about ready to buy another C.D."

Charge It

We live out here in the Boonies, in the Wisconsin
*   hills;*
The wildlife, the crops, the sunsets give us our
*   thrills.*

The clients I deal with pay prompt with check or
*   cash;*
They don't like to get bills with figures so rash.

Once a year we all load up and go to the Cities –
*   Minneapolis-St. Paul;*
The kids swim in the pool, we all have a ball.

Wherever we go that weekend, the financing is hard;
They are always asking me for my credit card.

When I whip out a big bill, the old Number Hundred,
They look at me like a bank I just plundered.

We go to the Mall, and buy something real funny;
The sign by the till says no bills over a Twenty.

You rich city dudes, your life styles so large;
You're lost in a pseudo-world of plastic charge.

So away we all head, to the Boonies and Hills,
And stack up some more one-hundred dollar bills.

Red, I figured out that you just want to marry me for my brains.

Yep. Best darn litter of pointers I've ever had, Clyde.

I think they have gone too far with genetic engineering this time, Clyde.

Dogs

Back when dairying was a family operation – mom, pop and the kids – there was always a dog, a family dog. He was part of the farm. He had a name and everybody knew his pedigree and where he or she came from. They knew where he slept and what he liked and disliked. When the dog died everybody mourned him, as he was part of the love giving and receiving of the family. He was properly buried and then talked about for years.

I always said hello to the dogs, and after a while, you could sense that they knew you. And, yes, veterinarians have a very distinct odor that follows them!

Now, later in life, I started to pick up on something, in that the dog slowly took on the characteristics of its owner. This started with the owners choosing of his choice in a dog to have, as likes tend to pick likes. For example, I had three clients that were bold, brassy, no tact, quick to disagree with you, in your face people. Guess what kind of dogs they all had? Large, in your face, barking German Shepherds or shepherd crosses with a demeanor that scared most insurance men. When you confronted them, each and every time you had to basically intimidate the dog to establish the pecking order. I had one that would challenge me every time. I would grab my grip and pail, head straight for the door and he would purposely get in my way. So, I would run into him with my knee or swing my grip or pail into him like 'Get out of my way!' That would be the end of it. I showed him! Until I was ready to leave. He would bark and his hair would stand up on the back of his neck. I would ignore him, get in my truck and leave. Many a person in town, who knew I did this guy's vet work would ask, "Doesn't that dog bite you? I'm scared to death of him and won't get out of my vehicle until Clyde comes out of the house or barn!" Dogs are experts at sensing fear and they know they are in control.

Border Collie owners are all sort of the same personality. Border Collies are timid, quiet, don't bark and always hide in the background. Some of my quietest, most reclusive dairymen had Border Collies.

Then there's the tail wagger. Oh, is he glad to see you come into the yard! He greets you and smells you, just a wagging the tail. That dog belongs to a positive, smiling, fun loving person. The owner will appear, "Hi, Doc! How goes it today? What a nice rain we had last night. Boy, the crops needed it!"

The forlorn dog doesn't come to meet you. He lays in the corner on some straw and if you look at him he sort of lowers his head and looks away. His attitude is like, "Don't hit me, please don't." He won't respond to petting. That owner is a negative person, probably a little rough with his livestock. Probably a yeller at his kids, wife and animals. You will never find a tail wagger on a negative's farm. I had a number of families with a lot of children – seven, eight or more. These were usually positive, hard working, well adjusted farm kids. They didn't have a dog, they had dogs! It seems like there had to be enough to go around. No birth control in the house, no spayed females outside either. A truism.

There were a few, but very few, farms that didn't have any dog at all. I would always notice when there was no dog that the eco-system was not quite complete.

It is a changing world, as our dairy world has gone two very diverse directions. There's the big mega dairies with lots of hired help and the owner more of a business manager and not quite so connected with his soils and cows. We've lost the dog. I think he got run over in the fast paced computer world. The sustainable, organic, small mom and pop operations I deal with in organics still have the family dog.

The only difference is he is now organic!

The Bet

There was an old vet by the name of Dr. Howe;
His favorite animal was a black and white cow.

One day while Doc was pulling a calf
The owner thought he would have a good laugh.

As he had checked the calf earlier, and felt a little
*   sack;*
"Ten dollars, hey Doc, he's a bull that will be black."

Doc knew immediately that he had a breech,
Took off his shirt so in he could reach.

"Okay," said Doc, "we'll see if you win."
And commenced to pull out a heifer, for he had a twin.

Do you think that the Good Doctor could just check
Snoopy's ears one last time, Clyde?

Sorry, Doc, but you are not on my list. I know its late but as long as you're here,
would you mind looking at Snoopy's ears for me?

Death

The one serious occasion when practice becomes deep,
Is that, in the Vet profession, we deal with delaying eternal sleep.

It may look like economics and cash flow to pay the rent,
But the gruffest old codger is usually full of sentiment.

Large animal practice is just like little pets, there is emotion.
When a good cow dies, there's a family commotion.

When you see an animal with death in its eyes,
They look to the Vet for a word to the wise.

"That animal has had a super place to live,
Here on this farm there was nothing you didn't give.

She had a nice stall, always clean and comfy to lay;
You balanced her ration and gave her top quality hay.

When that animal gets up to heaven and looks back,
She'll say – I had good owners, with Dorothy and Jack.

No, if you would have called sooner and lots of medicine I gave,
I'm afraid I wouldn't have stopped her from her bovine grave."

There's sure no cleavage left in these high producing cows.

Seed Corn

When I was a lad, we selected cobs of corn that looked especially good.
Now you need a degree for all the varieties to be understood.

In the spring when the corn is planted, the signs all appear;
There's Mallard, Trojan, Jacques and of course, Pioneer.

There are varieties for the silo, varieties for the white bag and one for the big blue tank;
Dairyland Seeds, Hughes Hybrids and a little kernel called Renk.

You read the literature of all the trials, and you're not any wiser;
You can plant Funks, DeKalb or genetics from Pfizer.

Stauffer is good, there's a yellow ear called Golden Harvest;
A Blue Blaney sign, plant Payco, they are all the best.

There's a local corn, that's got a following in this part;
A real good variety, we call it 'Seed by Carhart.'

In the fall when it's harvest time, it's all green and tall;
After the spraying, the cultivating, the rains and all.

All the ads, displays, trials, field tests and labor,
Don't mean a thing, cause you buy it from your neighbor.

Eternity

Let's talk about death and dying; that we don't
  understand.
Let's think about it, confront it, don't be like a bird
  with his head in the sand.

Now that I'm in my autumn years, I've sat by the bed
  and watched parents and close relatives go.
I've seen too many senile and wrinkled with a diaper
  on, as they don't know.

The worst feeling when one visits them and you sit in
  the chair,
Is that far off, way beyond look in their eye, that
  unknowing stare.

Then when they call you their uncle and
  congratulate you on your fame,
You just sit there in pity and realize they don't know
  your name.

You see in front of you a man of great dignity, pride;
  he had great success.
In come two nurses just to help the poor fella undress.

Why do we measure life with time and only consider
  the quantity?
Some people that have done nothing vegetate on into
  infinity.

The quality is what counts. Some people by the year
  forty have accomplished their mission,
While some people are on permanent intermission.

Now, one should not judge, maybe they have found
  peace and satisfaction in their own mind.
Until one walks in their shoes you don't know; we
  are not all one of a kind.

Returning to the aged and the health care facilities
  that lengthen the end,
I've seen a different side in the veterinary area, my
  friend.

The worst job in practice is to permanently take a
  creature's life, the end.
But when they are suffering and in pain, can't walk
  and have legs that won't bend,

Then we rationalize they are not human, we're doing
  them a favor to quit.
We euthanize them IV, so painless and quiet, but
  saddened a bit.

I always have a lump in my throat and a hidden tear
To know maybe this animal did want another minute,
  hour or year.

Does one get hardened with age because there's no
  tear in my eyes.
Every time I put an animal to sleep a little bit of me
  dies.

Then I try to relate to the human side and say well
  should we,
That would be a tough decision to lay on any one
  individual in society.

So don't let me be a burden on my kin, by having
  them come to see me as an obligation
When I'm senile and withered, on a permanent
  mental vacation.

Who knows, though, when I get there I may cling on
  and on to every precious minute
Just from drive and instinct to see what tomorrow
  has in it.

But for now I say, let me leave quick and fast like a
  bolt of lightning,
As this senility, incapability, not being in control, to
  me is frightening.

I stopped by there during the rut, and let me tell you, that's one frigid doe.

No blaze orange. No gun. He's hauling wood. Season's over.
Let's go bed down by his woodpile.

Can You Believe That?

I had a farm client that came from Switzerland and bought a really hilly, end of the valley, farm here in Trempealeau County. He milked cows and they grazed the hills just like back home. He was a very handy, jovial, positive fella that pretty much called for only emergency service.

One spring day, about the first week of May, he called with a cow down with milk fever in the pasture. This was pre 4-wheeler days and since we were both young, we struck out and walked to this milk fever. She was out in a little narrow valley way behind the barn. She was laying up against the fence in a little clump of brush next to the woods. I gave her a quick exam, tied her head around and started my calcium IV.

As we were quietly conversing, we were both looking across this narrow little valley and on the other side a nice big doe deer comes out of the woods into a very little clump of grass and brush, right next to the pasture fence and lays down. I was running the calcium in slow and as we were sitting in our clump and she was in her clump, she commences to strain. She, in about 90 seconds, delivers a little fawn deer right before our eyes. My client and I can't believe it. He whispers to me, "I don't think she even knows we're here."

She stood up and starts nuzzling the little fawn. The cow was responding well. I told my farmer friend, "Let's let the cow loose and not leave the way we came, but retreat up into the woods a ways and go around the point in the woods so we don't disturb her."

I stand up to take the halter off and she immediately looks up at us and freezes. She saw us right away. We retreated as planned and she never moved a muscle.

We both agreed that that was a once in a lifetime experience. No one I know has ever witnessed the birth of a deer in the wild. This was way before the motion cameras that are now so popular. As we walked home, my friend said, "Doc, can you believe that? Huh, can you believe that?"

Why large animal veterinarians make poor highway patrolmen.

Familiar Situations

You're a busy vet, day and night you cover a lot of
land;
No matter how old you get, there are some things
you'll never understand.

It's two in the morning, the farmer called after the
wedding dance;
"Hey, Doc, it's two a.m. but you forgot to zip up your
pants."

You're twenty-seven calls behind, you stop quick to
open a teat;
There stand the extra cows, with extra long feet.

"We finally got them in, Doc, one is blowing steam
from her nose;
I know you're busy, but would you clip off their toes."

The Early Bird, so cheerful, who calls at four a.m.;
"What are you doing, Doc? I've got one to clean
today if you can."

Then there's the guy who still milks at midnight;
"Want you after nine p.m., if that's all right."

"Please wait a while, don't come until after two
Because there's a good show today on Donahue."

It's Saturday afternoon, you're done on a hobby farm
and walking out the gate;
"Oh, my son's girlfriend's sister-in-law's nephew has
this horse to castrate."

An hour later you've castrated him against your will.
"Give me the slip. He lives in some suburb
someplace, I'll send him the bill."

"It doesn't matter, come any time today.
No one's going to be home anyway."

"I'll leave directions, one sick cow and one sick goat."
Twenty minutes of searching, you find his postage
stamp note.

The best client that never demands, and has a
smiling good look,
Is the one that's always home, along with his checkbook.

I have another good client, skinny little Fred;
His wife always gives me a loaf of home-made bread.

It takes all kinds, that's what makes the world go
round;
There's the telephone – you know I'd be on Welfare if
it wasn't for that sound.

Smile

If you are a grump, then you listen up;
Let's take that lemon, turn it into ade, and put it in a
cup.

Forget those bad deals, tough bounces and
unexplained falls;
Add some sugar, drink it down, sweeten those
stomach walls.

No matter how gloomy and dreary it seems,
Look through the fog, keep sight of your dreams.

Don't be sad, short-tempered and rude;
Ninety percent of success comes from your attitude.

All you're doing with bad humor, is a day getting wasted;
Put a smile on your face, let good cheer be tasted.

You see, I'm getting older, life is a very finite item;
Be a positive being, and join them – don't fight 'em.

Time is for all, for everyone the same;
So utilize those minutes, in eternity's game.

You should jump out of bed, with a smile and cheer;
But don't go too far with the song and the beer.

Negative people are static and status quo;
The positive folk are who make it go.

If your project fails, and falls on its face,
Pick yourself up, and get back into the race.

Do it with vigor and gusto, you've got what it takes;
Always profit and learn from those previous mistakes.

When I'm taken advantage of, by some darkened soul,
I'm glad I'm not like them, I keep track of my goal.

Set yourself a goal, make life a fun mission;
If you have a spouse, do it only with permission.

So, smile up, Padre, fill your vitals with cheer;
Make this a fruitful, happy and positive year.

No Bull

There's milking chores and young stock to be fed;
One of the toughest jobs is to get them all bred.

There's Repeat Breeders, Retained Placentas and
  cows draining;
Some have a tank, they went to school for A.I.
  training.

Silent heats, there must be something they need;
Call up the Mill. Let's sample and analyze the feed.

They already get the salt, trace elements, and a bag
  of vitamin;
My feed bills are so high it seems like a sin.

Hormones we use, Prostoglandin, Estrogen and
  Cystorlin;
Three weeks later, she's right back in heat again.

The vet, he does check her, with his arm he will
  palpate;
The uterus and ovaries, they feel just great.

The Technician will come, the semen he will thaw;
He'll breed her careful, with a little green straw.

We infuse, synchronize, and twice they are bred;
We'll glue on detectors, Ka-mars, they turn red.

They will get bred, on swelling and clear discharge,
  too;
Sometimes they are so quiet, they won't even say
  "moo."

But, the cows know what they want, their life isn't
  full;
The only thing they really need is a big, old bull.

Yes, dear, your stupid rabbits are being fed.

Guess Who

I was driving down the road, eating a beef stick;
In came a call from little Gerald Frick.

He's got a round little belly, and short little legs;
"Hey, Doc," he said, "I've got a cow without any
   eggs,

"She's ten months fresh, and I've never seen heat;
She's too good of a cow, to butcher for meat."

I put on a sleeve, so I could palpate;
To determine why her heat was so late.

"No problem here," I said with a smile;
She'll have a calf in a very short while.

"She's eight months along, give or take a day;
Dry her up and put her on hay."

"Not possible," said Gerald, trying not to laugh;
"We only had one Hereford bull, he was only a
   calf."

A month later, to the barn Gerald did race;
To see her deliver a little white face.

Wow! A lucky charm with a tatoo on it. That's really cool.

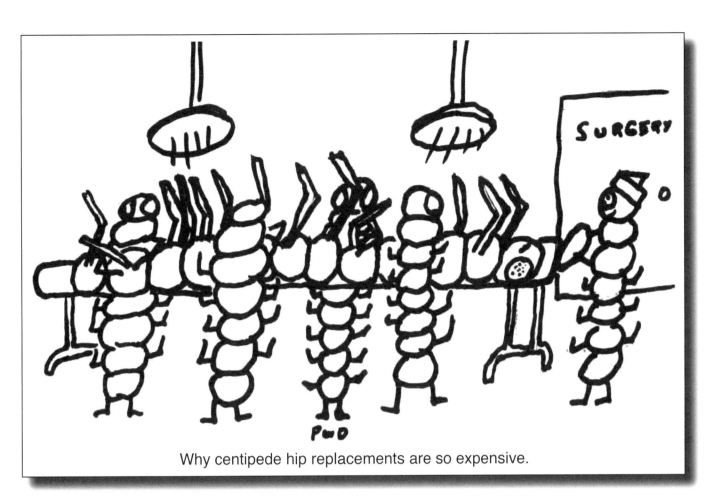

Why centipede hip replacements are so expensive.

Oh, look, I'll have butter
on my toast.

Oh, look, I'll have honey
on my toast.

Just my luck.
A dung beetle.

Success

Success is a state with a varied connotation;
Like recognizing your name on a dinner reservation.

It's not attained easily, it takes years of persistence;
It means going forward against all resistance.

Success cannot be inherited or willed to the heirs;
Cannot be store-bought or purchased at fairs.

Contributing factors include knowledge, wisdom and
    drive,
Keeping your mind straight while others take a
    mental dive.

The satisfaction is not when you're there on that
    final day,
But the pleasure of the journey, the fight along the
    way.

It takes much time, overnight success is insecure;
Minutes, hours and months must be invested to
    parlay a year.

So, I will keep trudging along on my journey, my
    friend;
Because my success is only partial, I'm not to the
    end.

Mechanic

Now, I can handle an oil change, and can even get
    the filter threaded;
But when it comes to the real deep mechanical stuff,
    I'm somewhat dumbheaded.

When they say, "Hand me the 5/16th, this 1/4 inch is
    too small,"
I just stand there, knee deep in sockets, and stare at
    the wall.

The real mechanics load up sockets and end
    wrenches when on a trip;
I fool them all, I just take one wrench – my vise grip.

I hate to see a mechanical pro come and check my
    fine work;
I have the edges all rounded off, sort of like a grease
    zirk!

I can fix a nose lead, I can even weld and make
    metal drip;
But, wherever I go, you know that I'll have my trusty
    vise grip.

I've seen big red cabinets – 17 drawers full, with a
    neat little rail;
My whole wrench inventory looks so alone, hanging
    on a nail.

When Christmas comes, I don't want all those new
    tools or a collapsible workbench;
Just go to the closest K-Mart, and pick up a cheap
    vise grip wrench!

So, Clyde, these are your new low fat milk producing cows you've been breeding for?

Weather Forecast For Lice

High winds tonight on back and neck areas. Very damp, humid conditions in scrotal area.

Operator, 911 please, this is an emergency.

King Kong

I was examining this cow, just talking along,
When I spotted this louse, my gosh, he looked like
  King-Kong.

I put down my stethascope and looked him in the
  eye.
He winked at me and said "Doc, you can kiss this
  cow good-bye."

He whistled out twice, did a little jig with a kick.
His progeny emerged four rows deep, and two thick.

I devour Ivomec and chemical insectides, I have no
  fear.
As you see, I'm a product of a Genetic Engineer.

My knees started knocking, my thighs, they got weak.
This is insane, could not even speak.

The old cow boss turned and looked at me with
  horror in her eyes.
Never were there ever so many critters and, oh, what
  a size.

The farmer, dejected, started, with head hung, to the
  house.
I went to my vehicle for my gallon jar of Super
  Louse.

I poured on old boss the garlic powder with the
  wonderful D.E.
I'll get those critters microscopically.

The garlic repels, the D.E. is infintessimally sharp.
In a matter of days, the critters will abandon the ark.

The engineers know their chemistry, you bet.
But they just ran into a biological vet.

Attitude

I have a client named Dale Thill;
In the middle of his yard sits an aeromotor windmill.

He's a friendly, jovial fella, that's polite and never
  rude;
I like people like that, with a positive attitude.

You can tell him news that's really bad;
He'll find something positive, and never get mad.

"It's a bad mastitis, you'll lose the quarter, Dale;"
As he's striping her out in a pail.

"Always was a kicker, and milked out hard;
Got a good heifer for her stall, she's going to the
  stockyard."

It seems like fellas like that have better luck;
He's got a little boy, just like dad, he's called Chuck.

The negative fella, who's always down in the dumps;
When it comes to losing cows, he sure takes his
  lumps.

One day Dale had one with a torn uterus;
He said "treat her best you can," he didn't fuss.

She cleaned, kept right on eating, didn't even bleed;
Sixty days later, she was all cleared up, ready to
  breed.

You treat them all alike, some live and some die;
Guys like Dale get a little extra, I don't know why.

They're polite, thoughtful, there's never any clash;
And usually, always pay their bills with cash.

I don't care how big and ugly she is, son. I'd marry her in a heartbeat.

Been down long?

Jake

There once was a man by the name of Jake,
He combed his hair with a steel toothed rake.

A huge burly specimen of a man,
Drank his beer from a garbage can.

Could eat an entire roast pig, head to tail,
Then pick his teeth with a ten penny nail.

He could grab a porcupine with his bare hand,
Jake was the toughest hombre in the land.

His voice was deep, gruff and mean,
Then one day he met sweet little Jean.

She cleaned him up, shaved off his beard,
She shaped him up, don't be so weird.

Now he's quiet and peaceful as a dove,
Cause big tough Jake fell in love.

Horns, Hooves And Tails

Ten thousand years ago, when cows were roaming
    wild;
They needed horns for defense, so they wouldn't be
    defiled.

They still have them yet, little buttons when they are
    born;
But by the time they are yearlings, they have a long
    sharp horn.

We burn them, use paste, gouge them and saw;
We remove a lot of them in the spring during the
    thaw.

You wonder if it's worth it, all that blood and gunk;
But it is better than abortions, from crowding at the
    bunk.

Or if you're nose leading a cow, and catch a horn in
    the gut;
Doesn't take long to freeze it, grab the saw and cut.

Now hooves, they get long and crooked and overgrow;
Man, does that hurt when they step on your toe.

They'll get a little infection, and get lame with foot
    rot;
You lift them up, trim and treat them with all the
    strength you've got.

A claw can get so bad, on the other one you put a
    block;
Then there's sole abscesses, from stepping on a rock.

There's one thing you watch for, when opening a teat;
You size her up, to see what she'll do with her feet.

We can cuss them, trim them, and do a lot of talk;
But without them, there would be no way they could
    walk.

Now, the thing that will really get you – never fail;
Is that long, wet, slimy, soaked up tail.

You're pregnancy checking a cow, with a beautiful
    udder;
And you get slapped on the neck, from a wet one in
    the gutter.

You can't cut them off, there would be moos and cries;
Cause in the summer, they need them to swat away
    the flies.

So, I just keep on tying them up, and giving them to
    farmers to hold;
Cause they need them in the summer, but not when
    it's cold.

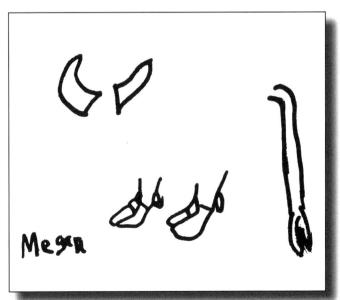

I want you to keep an eye on her for a few days.

Yes, sir! Jackpot! Don't you just love trophy hunters.

Hooves

"I've never seen a cow with so much pain, Doc. She won't get up and she just lays there and moans," the farmer said over the phone.

I'd been in practice 30 years and thought to myself, 'I bet he's got a cow with hardware, showing pain.'

When I got there, she was down in the stall. This was a stanchion barn and yes, she was in agonizing pain – lots of visible pain. I had never seen a cow express such pain. I gave her a physical while she was down, which is difficult. She had normal temperature, normal heart rate, normal uterus and udder – all fine. All systems normal. Omasal impactions will show pain, but I ruled that out also.

My diagnosis was open. I told my client I was baffled. I'd never seen anything like this before. I put her on a heavy dose of pain control and told my client he was going to have to help me with a diagnosis. I told him to call me with any new signs.

Well, two days went by and he call me with a new sign.

"What did you see Clyde?"

He said she had sloughed off all eight of her hooves and that she was in even worse pain.

"What? Your kidding!" I said. I'd never seen a cow slough hooves.

Sure enough, when I got there he had all eight hooves lined up on the walk like little soldiers.

A little light clicked on in my grey matter. I remember 'acute selenium toxicity' would slough hooves. I looked up in front, ahead of the manger, and here sat part of a 50 pound bag – wet, worn with the bottom busted out, and it was about half full, all sagged down.

I asked, "Say, Clyde, how long has that been sitting there?"

"Oh, a long, long time. I don't even know what it is anymore."

I got down on my knees and read "Selenium – Vitamin Pre-mix." The cow could just reach the bottom of that bag with her tongue, and took in about 25 pounds, probably in a very short time once she figured out how to get at it. I put her to sleep, poor thing was suffering so.

That time, I did give her a good physical, but didn't check out her environment.

Keep flying, Mildred. They say it is only 30 more miles by the crow to Hankee's.

Burp

It was the middle of April in Eighty-seven;
I was tooling along in my Veterinary Heaven.

When I got a call on my phone, "Come quick!
We've got a cow that is really sick."

It was a new client, a Mister McGee;
He came to the farm from the city, you see.

I drove into the yard, lickety bang;
In front of the milk house, stood the whole gang.

"What is your trouble that's an emergency type?"
Mrs. McGee was crying, a tear she did wipe.

"It's our best cow, Black No. 35;
She's down in the cornstalks, she's barely alive."

We hopped in my van, and away we sped;
Got out to the stubble, she looked nearly dead.

"Why, she's lying too far over, she can't burp up gas;
We've got to roll her over, so some air can pass.

You grab her head, I'll take a hind leg;"
We rolled her over, like beer in a keg.

"Ka-burp, Ka-belch," she said, her eyes they did
  blink;
The rumen was shrinking. Boy, did it stink.

"She'll be up in a short while, just give her some
  time."
When they die from bloat, they aren't worth a dime.

That rumen is a big fermentation tank;
That's what makes their breath smell so rank.

When they are on their back, they can't burp up gas;
They will bloat like a balloon, and away they will
  pass.

Odd Stuff

As a Country Vet, I'm amazed at the odd problems I
  discover;
I try to treat them all the best that I can, so they,
  hopefully, recover.

Have you ever seen a pet frog with a bad left hind
  leg?
Well, I have. He just kept circling this old wooden
  keg.

I checked him over, and found a broken left thigh;
I put on a little splint, now he jumps straight and
  high.

There was once a turtle, a genuine Snapper;
The owner brought him in, his lower jaw in a
  wrapper.

Gingerly checking his mandible, it was red as fire;
The poor old tortoise, had been run over by a tire.

I drilled four holes, and wired it with catgut;
Now watch out for your finger, when he snaps his
  jaw shut.

There was a female gerbil, by the name of Ruth;
Who came in with a broken lower front tooth.

The poor girl was starving, as she couldn't
  masticate;
With a little clipper, I cut the other one off, just like
  its mate.

'Cause a rodent's front tooth, I hope you know;
For their entire life continues to grow.

The last time I saw Ruth, she was doing just fine;
She just had a litter and was nursing all nine.

The oddest case, was a skunk that smelled like
  cologne;
He complained "All my friends, they leave me
  alone."

"No problem," I said, prescribing Lutefisk to eat.
Now the others are all playing at his feet.

The Worm

It was the summer of '99. I was out palpating away,
When this earthworm crawls up and says, "Hey,
  Doc, what is Y2K?"

"Well," I said, "Buddy, I'll tell it to you straight.
Us Homo sapiens may have just sealed your fate."

"Will free radicals invade our burrows and homes?
Will darkness descend over our soils and loams?

Please tell me why us earthworms should have such
  a terrible demise,
Simply because of you Homo sapiens' fast rise.

Your salt-base fertilizers make our soils like
  concrete.
We cannot escape, you know us poor earthworms
  have no feet.

Then you spray the environment, your arsenal is
  horrendous.
We poor creatures have no FDA or EPA to defend us.

Your agriculture today is not sustainable, no way.
Soils eroding, trace minerals depleting, just pour on
  more spray.

Look at us lowly creatures – we're rebuilding with
  fertilized castings aglow.
It's time you get more organic and biological and
  work with Mother Nature, you know.

Please no more Chernobyl, Three-Mile Island or
  Love Canal today,
Get your act together and figure out Y2K.

We were here before you in the Triassic, the Annelid
  line.
You're a Johnny-come-lately, you Hominids, in
  geological time.

So leave us in peace while you destroy your
  existence,
We'll wait for the aliens and their new subsistence."

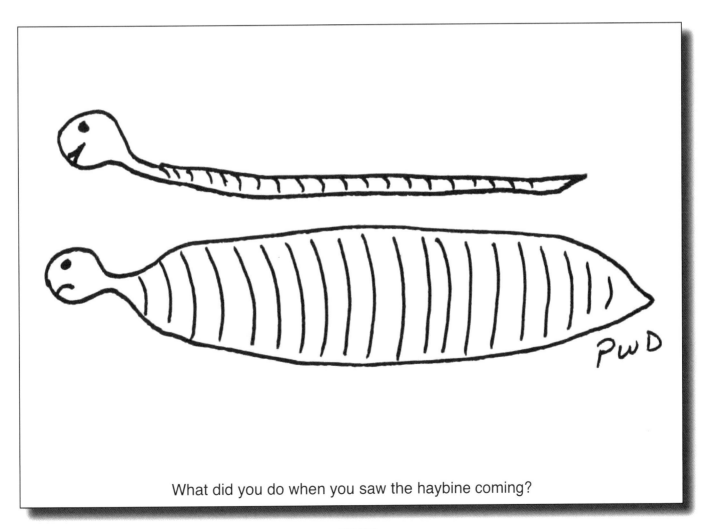

What did you do when you saw the haybine coming?

95

Snake

I once saw a snake with glasses.
He had trouble crawling through grasses.

To crawl into his hole took him quite a while;
Such a handicap, to be a nearsighted reptile.

Then one day, he was talking to some Rattler
  friends;
They told him he should get a pair of contact lens.

Now, his eyes don't fog up in his hole when he
  sleeps,
He can catch a Kangaroo Mouse before he leaps.

He's so happy to be living in this technological age;
With all of the female snakes, he's now a
  Herpetology rage!

Hey, we gotta keep this thing! I rattled it twice, and WOW, did they clear out!

Me, Doc Dettloff, going into the barn in the 1980s.

Watch the mud hole, Doc.

I don't think the boss knows what separating means.

Rufus Quillan

I had a grumpy, old client, name Rufus Quillan;
Everything sick got two shots of penicillin.

One day he called with a Jersey a-dying;
I went out to look, he wasn't a-lying.

Put her on forty cc's of Oxytet;
In a few days she will be better, I bet.

That will be twenty bucks, the medicine and call;
He paid me in cash, his money rolled in a ball.

Ten days later he called me back, "She can't even drink;
She's teetering on death, she's right on the brink."

I checked her eyes, ears, udder and nose.
"What little she eats, out the back it blows."

Another one with severe anemia;
This poor girl is dying from leukemia.

"She's a dyer, you say, not worth a dime;
If another one gets sick, I won't call you next time."

Now when I drive by, I simply wave at Mr. Quillan;
Standing in the barn with his bottle of penicillin.

Milk Haulers

When I was a farm boy a sight commonly seen
Was a canvas aproned burly man picking up your cream.

We stored it in ten-gallon cans, in a stock tank with ice it was stuck.
He'd grab those heavy cans and throw them in some old truck.

Today's scene is a young man dressed in white, fingernails cleaned to the hilt,
Drives up to your milk house with a stainless steel tank on a Peterbilt.

He's quiet and nice and calculates your milk from the measure,
Says hello, good day, thank you, ma'am; he's a real pleasure.

They all have some clever name for their vehicle on the bug reflector,
But, I have yet to see one that calls himself the 'milk collector.'

There's 'white lightning,' 'udder cola,' 'barn stormer,' 'big red' and 'udder delight.'
They pick up milk all day long and unload it at night.

They are ambitious and clean and seldom very late,
They are not paid by the hour, but by the hundred weight.

The cab has a stereo radio, warm heat and cool air,
And a soft, swiveling, spring-loaded adjustable chair.

When they drive away, they wave and smile to please.
They are taking the milk to be made into cheese.

When they park by the barn all filled with hay,
Be sure and stay out of the milk hauler's way.

Bo-Peep

Little Bo-Peep,
She lost her sheep.

She called her attorney, Mr. Denet.
Said, "Let's sue the incompetent vet!"

"Now," he asked, "Why should we sue?
What wrong did this vet do?"

Bo-Peep replied, "His diagnosis was wrong.
It was such a big technical song!

He talked and talked on into the night,
About a tapeworm parasite.

He made me so irate!
I don't feed me sheep tape!"

The jury saw the evidence submitted.
The country vet was quickly acquitted.

The jurors just couldn't believe
The vet had the tapeworms in a sleeve!

I've never known such an avid collector.

What is this world coming to in our schools? No basics any more, Peter.

No, I'm sorry. Attorney Bagley, our founder, can't see you.
He was killed in the civil war at the Battle of Chancellorsville.
Mr. Cooper is deceased. Attorney Helms died of TB.
The Honorable Mr. Smith is dead.
Wilson retired when Eisenhower took office, but Dudley can see you.

Certainly I'll loan you $500 on your eight head of fresh hiefers.
Just come on in.

Money And Life

So many people today have a problem with money;
When the checkbook is empty and the rent due, it's
    really not funny.

Here are a few tips to young folks, just little ones,
    not large;
First thing you do is throw away the plastic that you
    charge.

Then there's a new word you have to learn and it's
    called Budget;
The debits and credits do work, and don't try to
    fudge it.

A principle a lot of people never learn or even heed,
Is the difference between a want and a need.

When young, and money short, with a number of
    mouths to feed;
Just ask yourself, is this something I want, or
    something I really need.

No matter how tight the budget, or how much you
    have to pay,
Always keep a little back for that emergency or rainy
    day.

When you're older and have a little extra free cash,
Divide your money into serious and fun, which you
    can stash.

Serious dollars all go for principle reduction, IRAs
    and CDs, all increase net worth,
Fun money is for the wants to fulfill, the fun and
    pleasures of earth.

The fun money is the unexpected refund, the gift, or
    the sale of firewood,
Then keep that in big bills really handy, it makes one
    feel good.

After the debt is removed and all the liabilities are
    Accounts Payable,
Then only buy big items when you can lay cash on
    the table.

Don't try to keep up with the Joneses, or Mr.
    McDougal,
Work hard when you're young, save and be a little
    frugal.

Don't get lavish when young and have a car with a
    neon windshield wiper;
If you're pulling your own, you have to first pay the
    piper.

Don't be envious of the kids with the silver spoon,
They never had to play a laboring tune.

To inherit a large amount makes your pride all for
    the worst.
You're not a man of your own as your parents have
    your life all rehearsed.

To stand on your own and maybe not have a lot,
Is still better because your dignity and self-esteem
    you've certainly got.

Separators Acquired

I went to an auction in mid-January;
Bought a red separator, the label said IHC.

It came with two strainers, a round rubber ring;
Had an extra crank, an odd looking thing.

Cleaned and hung up the strainers, hung them in the
    fourth row;
The handle looked familiar; where it fit I didn't know.

I looked at my extras, and right in front of my nose,
The separator missing its crank was a No. 3
    Primrose.

One day in June I had a vet call at Clarence LeFever;
He had a cow down in the pasture, real bad with
    milk fever.

I put the nose leader on her and was running in
    calcium,
Looked over on his junk pile, there laid a Melotte
    from Belgium.

"I'll trade you this call for that Melotte, Clarence."
"That's a deal, Doc, it belonged to my parents."

Continued on next page

Continued from previous page

Got it home in my shop, apart I took it;
Cleaned and washed it all up, so it would all fit.

It's sitting in Row 5, all original and complete;
Salvaged from a junker's torch, all that pounding
*   and heat.*

I have a burly old junker friend, who swears much
*   and is real gruff;*
He saves me separators, separator parts and chews
*   Copenhagen snuff.*

A week ago Wednesday, he stopped at my farm and
*   spit out his chew;*
"Hey, Doc," he said, "I need twenty bucks for this
*   dandy Royal Blue.*

I got a heifer not eating, looks like for living she
*   don't care."*
Sure enough, the next day I treated her for hardware.

"Where'd she get hardware?" he barked, looking
*   ugly and mean*
As he leaned up against a rusty old machine.

"Beats me," I said as I stepped around the wire and
*   nails.*
We walked to my car between rusty barrels and
*   leaky old pails.*

Then there's August Kabuna, my tightwad old friend;
At auctions he buys these fifty-cent boxes and stuff
*   he can mend.*

He brings me Alpha discs, tanks and little round
*   floats.*
He's kept one tabletop separator; says he going to
*   milk goats.*

Last September I delivered a calf, on a late night call,
For the son of an old client, name of William Westphal.

"Hey, Doc," he chuckled, "thanks for helping my son.
I'm giving you this old separator made in Port
*   Huron."*

Separator Collector's Poem

I started a neat hobby, collecting separators in '74.
Now I've got over 200, and looking for more.

Being a Minnesota farm boy, I sat on many a milk stool,
Stripping out the ole boss, thinking I was real cool.

The education I got is valuable yet,
'Cause now I'm a Wisconsin Dairy Vet.

Remember the wet tail, slapped across the face;
I learned some new words, and they weren't dinner
*   grace.*

How about Old Nell, with the stepped on tit;
My dad would grab a Naylors and dilate it.

Our first separator was heavy and red, the name I forgot;
Oh, yes! I remember, a Belgium Melotte.

That funny hanging bowl, a spinning away;
I remember my mom washing it every day.

Then Dad went to an auction and bought a De Laval 17;
We were so proud; it was so shiny and clean.

Our one neighbor was a conservative, with his old
*   Primrose;*

They had more cows than us, but that's how it goes.

Another neighbor, a fancy young dasher,
He went broke; they had a McCormick self-washer.

Now my collection boasts a Galloway, Iowa, Lacta
*   and Viking;*
But my big Diablo, is most unique and striking.

Last deer season I bought an American-made, in
*   Bainbridge, New York.*
I purchased it from a neat old client named Anthony
*   Bork.*

Today's a bright day, even though it's rainy and mean;
I have an auction bid in on a De Laval 618.

The tabletop models – Montgomery Ward, New
*   Prima and Royal Blue;*
They are getting expensive. Oh! What can I do!

I just keep looking at auctions and sales;
So when I talk to a fellow collector I can spin some
*   tales.*

The memories I have of the separators I did crank
Can only come from a farm, I do thank!

104

One Hundred And Twenty Years

If the pioneers could see this country now, when
  their covered wagons were getting stuck;
Now we have freeways that can move anything in a
  truck.

The logs they cut by hand, with two men making
  notches;
Now we log the woods with a chainsaw, and one
  man watches.

Manifest Destiny would have been very obsolete;
With our modern machinery at their feet.

What they ate, they raised, and you can believe little
  was wasted;
With salt and spices, food was seasoned, and good it
  tasted.

Now a fax machine transmits information in
  seconds, of course;
The Pony Express wouldn't have needed a horse.

We go coast-to-coast in jets, with flight attendants to
  help;
Our forefathers alertly carried guns, to save their
  scalp.

I had a great-great-grandfather that lived in a sod
  house with worms and bugs;
Now we have central air, antennas on the roof and
  Monsanto rugs.

The old-timers worried about locusts, grasshoppers
  and slugs,
Now our biggest concern is that our kids stay away
  from drugs.

Imagine, in only 120 years, the change in our society
  and culture;
We have no comprehension of what's ahead for our
  offspring in the future.

When we get to the extinct hominid strata in our excavation, students, rubber
gloves and masks are required, as this layer is still loaded with toxic pesticides,
insecticides and herbicides that can harm some of us.

Do They Know?

Farming was a man tilling the soil and raising
  livestock in solitude.
Now we have an era of government programs and the
  Big Brother attitude.

There're price supports, check offs, set aside land,
  minimum tillage and C.R.P.
Now part of each farmer's time is spent in the
  courthouse signing up, you see.

The barnyard runoff and cattle in streams raise the
  bureaucrat's hair.
Those navigable streams belong to the D.N.R. and the
  Corps of Engineers to share.

I know we want to preserve the land and resources
  and soil,
But we still have to utilize what is there for our toil.

Milk and meat residues must have a zero tolerance
  anymore,
The consumer does not want anything tainted at the
  store.

The FDA can find one part per billion in an animal's
  liver,
While big cities can dump raw sewage in a river.

Rats get cancer when fed at one hundred times the
  normal rate,
So production is stopped, we'll all die, it's too late.

Have we become alarmist, is our society worrying too
  much?
Or are the decisions being made by people out of touch?

Maybe, I don't know, we are all ignorant and need
  protection
And a few wise ones are helping save the masses from
  extinction.

When I look way back and see we were covered by a
  sheet of ice
Only fifteen thousand years ago, the fourth glacier
  wasn't nice.

It didn't respect erosion, topsoil or stream banks,
It just bullied its way over all life. No thanks.

So when glacier five cools us off and the bureaucrats
  vote twice
They will be vetoed by Mother Nature's sheet of ice.

Or if the greenhouse effect turns us into a tropical
  paradise,
We'll not raise corn, alfalfa or oak trees; instead it'll
  be rice.

So don't get up tight about the farm programs or parts
  per billion,
The dinosaurs roamed my farm an age ago, about 65
  million.

Time is so large, our minds so very small.
Just enjoy our beautiful transient spring and fall.

Dear Midwest Bioag: My soil is terribly unbalanced and I am losing my coconut crop.
Please send a soils consultant as soon as possible.

Yep, Clyde, first it was BGH, then by-pass rumen protien,
now they bypass the whole dern tootin' match.

Craig

There once was a little calf by the name of Craig,
Who picked up two lice one day, off his mother's bag.

At first they didn't bother a lot,
A little bite and a rub, one little spot.

Then the second generation by the hundreds, cut a
  new trail,
He itched so bad he lost his hair by his tail.

By the time the third generation of lice, he was
  having a fit,
With a very low hemoglobin and lower hematocrit.

The owner looked into the pen and was about to brag,
Until his eyes feasted on poor, anemic, little Craig.

In came the vet with his drugs for infection,
Gave Craig a physical and a thorough inspection.

No fever, all systems running real nice,
Only problem is, Craig's called home by ten
  thousand lice.

He's anemic, itchy and has a look of sorrow,
Let's rid him of lice and help his bone marrow.

Iron and vitamins, both a shot in the neck,
And for lice, we'll give him some Ivomec.

In three days, Craig looked much more hip,
Probably because all his boarders had jumped ship.

Now he's healthy and happy, a handsome young stud,
No longer anemic with thin, tired blood.

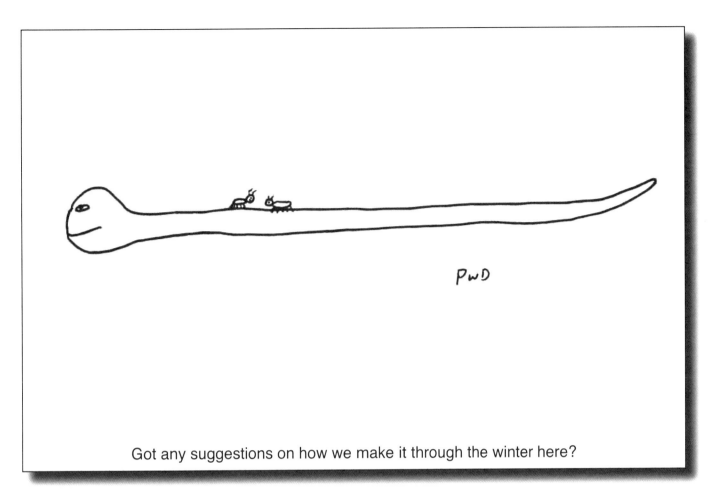

Got any suggestions on how we make it through the winter here?

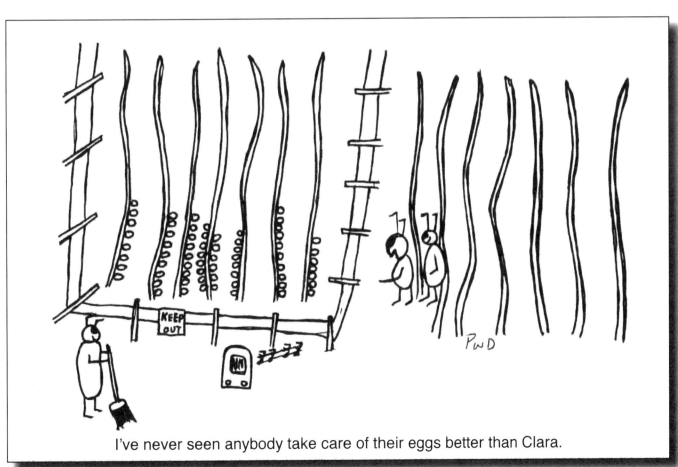

I've never seen anybody take care of their eggs better than Clara.

OB

There's one call that I receive, that is like music to
me;
It's what's known in Veterinary Circles, as an OB.

Every OB is a new challenge that gives the Vet self-
worth;
To assist an animal that is having difficulty with
birth.

I've delivered calves in hay lofts, pastures and
creeks;
Experience is a valuable tool, I've learned lots of
tricks.

I've delivered calves in mud, rain, snow, sleet and
next to waterfalls;
They've been in ditches, gutters, under machinery
and crosswise in stalls.

It's not uncommon to have the head back, looking in
the wrong direction;
When one grabs the lip and jaw, watch those sharp
teeth when you make the correction.

A real clue for twins, is when the tail comes first –
that's called a breech;
To flip those legs back, you're up to your armpit with
your reach.

A tough one that wears you out, until you want to
cuss;
That's when you've got a 180° torsion of the uterus.

One reaches in, grabs the neck, and rocks it – not
poking the eyes;
Remembering that a great percentage of them are
twisted counterclockwise.

When they are delivered, you knew that you would
win;
You always, always, reach in, to check for tears and
a twin.

When they are delivered dead, I tell the owner never
look back;
But, he'll always lift the leg to check for the teats or
the sack.

When all the tricks fail, and you get that feeling of
rejection;
One just goes out and gets his surgery tools, and
does a C-section.

I often wonder what thoughts go through that little
calf's brain;
When I first put my hand into his world with that
shiny OB chain.

When I hook up around his legs, he'll always pull
back;
But out he will come with my trusty Calf Jack.

Mother Nature then takes over, with her protective
purse;
Within an hour, the calf will be up and will nurse.

Kids

Farm children are a unique group all to themselves;
I've got them studied, categorized and put on a shelf.

There's the timid group, that bows their head, so shy;
No rattling of the tongue or always asking why.

Another group that's observant and a little more alive,
I visit with them and when I leave they give me 'five.'

The beggar group, they're sort of like a young bum;
They always have their hand out 'cause I give them
gum.

There's a big-eyed group, I guess I'm their idol,
Their parents call me "Doc," they're impressed with
the title.

A lot of farm kids are real workers on the job,
I walk by while they're milking and say, "Good
morning, Bob."

Then there's the helper – the little grip grabber,
Assisting me on my call, but with a constant jabber.

As a group, farm kids are extremely polite.
They usually say "Have a good day," "Good
morning" or "Good night."

There is a smaller group that there is no comparing;
They are the ones that are totally overbearing.

When I went off to college, I was hungry, naive, and
had no charm;
I got a part-time job, no problem, 'cause I was from
the farm.

110

Give her one more crank, Clyde, I think it will come.

Just let them be. It's teaching them their eye-hand motor functions.

Farm Boys

I've seen many a lad that was born and raised on a
   dairy farm;
They're usually quiet, pretty quick mind, resilient
   back and a strong arm.

They can deliver calves in the night, with tangled up
   feet;
They can detect a hot mastitis with one strip on the
   teat.

When fall approaches, they will have traps in the
   creek,
They can put out a fox set, they know every trick.

All of them, at one time, planted many a tree;
To watch it grow is rewarding, it makes one feel free.

The different trees they know, not by looking at the
   leaves, but the bark.
They know which one a raccoon will go up, when it's
   treed in the dark.

Broken metal, oh, yes! They can weld it together;
And put it back on a shaft and it will wear like leather.

That small engine they'll overhaul, put in a new
   piston, take off the head.
When they are done, they will have it looking new,
   painted New Holland red.

With a sharp knife they'll butcher a pig, cow or steer,
But their biggest thrill of all is field dressing their
   first deer!

The high school coaches like them on the athletic
   field, wrestling mat or court;
They give a supreme effort in each and every sport.

Some excellent ones stay on the farm; some excellent
   ones go off to college.
They set out on their quest, to seek out new
   challenges and knowledge.

As I look upon the group, what do I see?
One works in a big city bank, he's an Executive V.P.

One is a Nuclear Engineer, working with atoms and
   little things we don't know;
I hope he's got it figured right, so we don't disappear
   in a blow.

Some are computer experts, and talk in a language
   I've never heard.
I thought language was nouns, adverbs and proper
   use of a word.

One flies over my house weekly, he's a commercial
   pilot for TWA,
He married a California girl and they both live in L.A.

It all comes back to the work and responsibility of
   being reared on a farm;
The soft life of cartoons, electronic games and VCRs
   can only do us harm.

Let's dine at the farm across the road tonight. I hear it's filthy over there.

Kids

While in veterinary school, we had an old seasoned veterinarian come back to teach large animal medicine after being in practice for about 30 years. When he spoke it was with wisdom. He had been in the trenches and had walked the walk. He also gave us a little practice philosophy. He taught us that it is really important to pay attention to the kids. Learn their names, give them a little job and makes them feel important. I took that lesson to heart.

I started practice in July of 1967, and one of my first calls was on a farm that had four boys, all quite close together. When I got to the end of their driveway, I wrote all four names down in order - oldest to youngest. I put their initials as an acronym above my visor and memorized their names. I had all kinds of slips up there for many years.

Next time I hit the farm, the second oldest was there and the youngest. As I got out of my vehicle, I said, "Hello Eddy. Could I get you to help me by getting me some water? Hey Josh, would you carry my nose lead and show me what your dad needs done today?" You could just see those kids beam with pride because Dr. Dettloff knew them.

Another thing to do is to complement them. Such as – Where did you get biceps or muscles like that? Man, you must be strong. Where can I buy muscles like that?

When you won the kids over, you've won the parents. It's as simple as that. If the parents hear you, all the better as everyone is proud of their offspring. Another thing I would do is to carry Wrigley's Spearmint Gum – you know when you could get five sticks for a dime and when everything didn't have to be sugar free. I would make a little game of it, especially to the younger crowd. I would kind of sneak it to them by looking both ways and going Shhh...with my finger, like we were doing something secretive. Or sometimes a wink would be as effective. A big one eyed wink meant it was just between them and me and nobody else had to know.

I had one little 5-year-old boy that asked me if I had any gum to give him. I did but had left it in the truck so I made amends and gave him a stick when I left. The next time I was there, I wasn't in the barn 5 minutes and he said, "Mom says I'm not supposed to ask you for gum anymore. That's not polite." I said, "Oh, okay." I went out to get some fluids for an IV and grabbed some Spearmint Gum and as I was treating the cow in the neck, I slipped him a piece of gum with a wink. He smiled and stuffed the gum in his mouth. About the time I was finishing up, his mother appeared to feed the calves. As soon as she got close to him, he pipes up, "I didn't ask for the gum, mom, Doc just gave it to me, right Doc?"

One of the joys of practice was to watch these little farm boys and girls grow up to play sports, be in 4-H, FFA, go to college and turn into wholesome men and women with deep wonderful roots.

Junior buddied up with some fluke friends and they're waiting for a ride to the liver.

The Bee

When I had my multiple man veterinary clinic, we hired a high school boy to be our guy Friday to help us at our clinic. We really lucked out. This lad was from a large family and was one super worker. He did everything on a run, was thorough, didn't mess up and had a positive fun-loving attitude. It didn't take us vets long before he was a favorite with all of us.

When he was on vacation from school or had a day off, we would take him along on calls just for the company. During the summer he worked more hours as we would all come up with personal jobs for him to do.

One warm summer day, he was with me and we had the windows open in my pickup, and a bumble bee entered the cab and was buzzing around. We both commenced to wave our arms and get the bee to leave the cab. No more bee. We looked around, waited – no bee. Wow, he was gone. We rolled up our windows and drive along.

We go about six miles. We're visiting and enjoying the scenery. He turns his head to look out the side window. I pick up a wooden pencil and just touch the back of his neck and say "Pssst".

He yells, hit's the roof, slaps his neck, thinking he just got stung.

I felt a little bad for doing it, but it was so much fun, I would do it again if I had the chance.

Myrtle, your ovipositor is dragging.

Rain Gauges

Farmers watch the weather closely. Nearly everyone had a rain gauge and it was usually someplace between the house and the barn. I noticed them whenever it rained.

"How much rain did you get, Clyde?"

It was either .4 of an inch, or .6 of an inch, they knew the exact amount; or they admitted they hadn't checked it yet.

Just to put a little excitement in my life, if no one was around when I arrived, I'd spike their rain gauge by adding about an inch of water. If they hadn't checked it yet, that would be the topic of conversation as I was going to my vehicle to write up the bill and get paid or to leave. We'd start talking about the rain last night and the farmer just had to go check how much rain he'd gotten the night before.

"An inch and 3 tenths. Wow! I didn't hardly think it had rained – it didn't seem that wet!" he'd exclaim.

"Well, I know we had quite a shower and it was sure wet at my place," I reply. Then I'd leave.

I pulled this for many years. Some farmers caught on, but it was always fun to see their initial reaction.

Pasture Consultant Vet

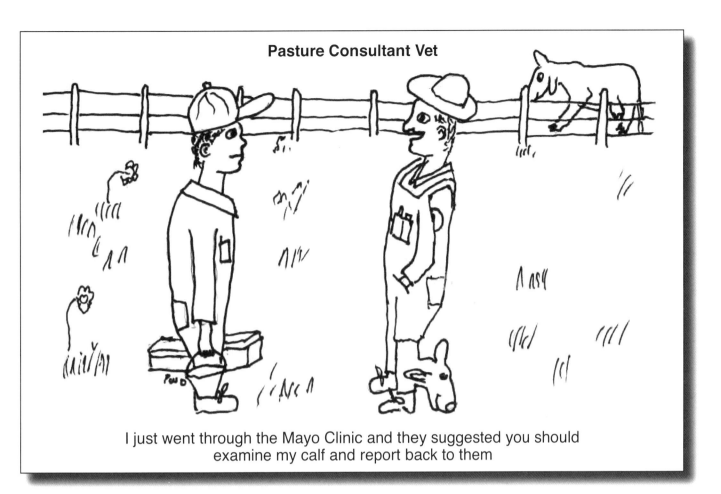

I just went through the Mayo Clinic and they suggested you should examine my calf and report back to them

You just wait. They will be back milking after they kill all those poor buggers off.

Hermit

I once knew a hermit named Dominic.
He never got lonely, he never got sick.

His house was ancient, made of logs,
He had lots of cats, but he had no dogs.

He cut a few ties, had some cows that were beef.
Paid minimal taxes, but took no relief.

He was a nature man, close to Mother Earth,
The plants and animals were observed with spring's
    annual birth.

His ear was tuned to crickets, east winds and tree
    frogs,
He talked to the woods while he sawed his oak logs.

No need for plumbing or lights, he had no electricity.
He had a small list of what he called a necessity.

He could barely read, he was not educated by any
    means,
His wisdom came from common sense and logic
    that's in the genes.

No rock on the radio, or Donahue on TV,
He was more interested in harvesting honey from a
    bee.

His mind was uncluttered, his thinking simple and
    clean.
He had an old Ford pickup, called it his driving
    machine.

I attended his funeral at the Mortuary – no church
    with a steeple.
I was the last to arrive, to make a total of twelve
    people.

His life was full; his simple journey was complete.
Everybody doesn't need all those material items at
    their feet.

Why?

Have you ever wondered why they say "Bless you"
    when you sneeze?
Everybody knows that's not a fatal disease.

There's no temperature rise when a cow gets Milk
    Fever;
They get Calcium IV to correct and relieve her.

It's called pinkeye when the cornea turns white;
If it gets bad enough, you endanger their sight.

Dystocia is one of the worst sounding words on the
    earth;
Why, that's a natural event when a cow gives birth.

That teat that's plugged with scar tissue inside her;
It never has looked anything like a spider.

Femur bone truly sounds like a bone in a leg;
But humerus should be a smiley face on an egg!

Lips, nose, teeth and ears all sound right,
But I would never use pupil when it comes to sight.

When the statement is sent, it's no big thrill,
But, why isn't it a Bob instead of a Bill?

Tony

I have a dairy account I enjoy that is very unique,
For he will drive two hundred miles to look at an
    antique.

Fishing lures, oak dressers, an old toy pony,
He's very knowledgeable on all; his name is Tony.

An old spinning wheel, hand blown fruit jar, old
    wooden fan;
The real good stuff stays in the house close to his
    wife, Diane.

Their garage no car enters, in front it must stop;
As you see, they have converted it to an antique
    shop.

About the only thing he's never owned is a meteorite
    from a crater,
But, he'll have one soon, if he keeps reading the
    Antique Trader.

I'll have to admit, he's gotten a bunch of my cash.
I recently bought a knife rest, old scale and two
    grass seeders to stash.

It's an enjoyable vet call, he's never a crank.
There will often be a unique item sitting on the bulk
    tank.

They do well with the cows; antiques provide a
    mental rest.
It's much better than being dairy cow possessed.

Tony always has a smile, and walks through life
    satisfied.
He'll never be in a mental institution all tied.

His philosophy is to get something good out of each
    day,
And hopefully, pick up a rare antique along the way.

If you really are the other white meat, show me your bacon, please.

Clarence

There once was a timid bull named Clarence;
Who was brought to be one-half of the parents.

"I'm worried, Doc, I never see that boy breed."
We looked over the fence, and he was eating feed.

"Let's go in the barn and see what he's done.
Yes, number 7 is pregnant, so is 12, also old 21."

"What do you know," said the farmer so proud.
"Here I thought that Clarence had his head in a cloud."

"No," I said. "He's just timid and shy;
But, it looks to me like he's clever and sly!"

"Well," said the owner, thinking it over a little bit more,
"I guess he's like me. When I go to bed I always
    close the door!"

The spider's got me! The spider's got me!

Now children, wash your pincers before lunch.

Spider in the teat is just a figure of speech, so
don't panic, Clyde, over some little ole barn spider.

Bad Habits

Do you have any bad habits you want to admit?
Well, I've got some things that give my poor wife a fit.

A bad one, I guess, it must be real rash,
Is my leaving my billfold on my vehicle's dash.

I used to always lose it – pulling calves, fixing cows,
    even pruning trees.
Why, once it showed up inside a client's deep-freeze.

I've often experienced the panic of the empty back
    pocket syndrome;
Applying for a new license, Master Card and no
    numbers to phone.

Only problem now, is when I take my wife out for
    dinner and reach for cash;
My money laden billfold, yes, sir! It's home on the dash.

The worst one of all, and to break it is real tough,
I always carry that round can of Copenhagen snuff.

Some men can put in a pinch and for hours they will sit;
But, I'm a salivating clown, 'cause I have to spit.

A nice white shirt, so fashionable, not gaudy, just
    plain;
You look down and notice that little brown stain.

Now, I'm not a boozer, I gave up being a drinker;
But in my garage, I just love to tinker.

I go in the house, wife's got it spotless and clean;
I'm followed by a strong odor of grease, oil and
    gasoline.

I can memorize drug dosages and phone numbers I
    call,
But I can forget to pick up the boys after practicing
    football.

I've driven past the farm I was headed to, waved and
    honked like a jerk;
Turned around, went back, only to see the farmer
    stand there with a smirk.

But, one thing in life that is very necessary
Is never, never forget your wedding anniversary.

Looks like Elsie got into the hemp patch again.

Snuff

Did you hear about the Hereford that chewed
   Copenhagen?
When she got up in the morning her tail was a
   draggin'.

One small pinch between her gum and lip
Would lift that tail right up to her hip.

Long leaf tobacco would do, it comes in a pouch;
A very happy Hereford, she was never a grouch.

When all her sources would run real low,
She would hit the pasture for wild tobacco.

She had a little calf, she called him Skoal;
But her morning tobacco, that was her goal.

It would get swallowed down into the rumen,
Past the pharynx, larynx, sliding down the lumen.

Everyday it was the very same routing;
She had to have her morning fix of nicotine.

There was just no way she could get enough
Of the tobacco plant product we all call snuff.

Then one day she was listless, her appetite off;
A short time thereafter she developed a cough.

The vet gave her a physical to find the answer;
"This poor old Hereford's gonna die from cancer."

120

Cliff

Cliff had good cows, he got up with the sun;
They really produced, he daily sold near two ton.

About twice a year, on a sunny winter day;
I would diagnose a cow with a left side D.A.

The surgery is routine, Clifford, but you'll have to
　pay some;
To get me to sew down her displaced abomasum.

Now Clifford is frugal, had a nice wife and daughter;
But his surgery cows always went to slaughter.

There's no cow in my barn that you'll ever cut;
And reach in with your arm and sew down her gut.

Then Clifford got sick, and his spirits got bogged;
When he learned that the plaques had his arteries
　clogged.

A conference was held with his wife and young lass;
The doctor said to live, he needed a triple bypass.

A month later, he called fit as a fiddle;
All healed up and home from the hospital.

"You're looking good, Cliff." "I'm fine," he did chide;
Then I told him "We've got an abomasums on the
　left side."

I closed up my grip, and headed towards my car;
When the words he spoke set my mind ajar.

"Let's do surgery," he said, "that girl needs a chance."
I stood by the door and nearly wet my pants.

"Surgery you say, why no problem," I told Cliff;
"I'll get my knife and we'll do her in a jiff."

He talked of medical progress and spun me a yarn;
"I never thought I'd have surgery done in my barn.

"She's a nice looking cow, don't cut her too far;
"I sure don't want her having a scar."

Now the cow's milking fine, and Clifford looks good;
Sometimes life takes a turn for the good.

What Happened To Your Dad's Cow?

I had a client just outside my circle of practice that would call me about every 6 months. It was always an odd call, completely out of the routine. At first I thought he called me as second choice until I realized that he either ignored the normal, routine calls; or dealt with the routine calls himself. He was the kind of man that I sensed was quite dominant – totally ruled the roost. One that always kept everybody at arms length mentally. I had a business relationship with him as a veterinarian and it could go no further.

One day in mid-summer, he called with a cow with a leg injury. I got there late morning and found one lone cow in the barn. When I drove in, I noticed someone in the distant field mowing hay with a tractor and haybine, so knew I'd be treating the cow by myself. I got my warm water from the milk house, added my disinfectant and headed to the cow. As I approached her, I saw her left hind leg and stopped in my tracks. The quad muscle on the back of the thigh, that makes up the big ham-string muscle, had a huge area missing skin and a lot of muscle gone. I set my pail down and start to look her over. I take her temperature and wonder what in the world happened. About that time, a little farm boy about 7 or 8 years old, comes into the barn and wanders up to me.

After looking her over good, I inspect her leg closely and find a nice round hole in the skin on the inside of the thigh. Being a deer hunter, I look closely again. The hole in the skin looked to be about the size of a 12 gauge slug hole and the exit wound looked just like the damage that a slug would do blowing the outside of the leg away.

I quietly ask my little friend, "What's your name?"

I get a one word answer. "Dennis." (The apple didn't fall very far from the tree, you know.)

I then quietly ask, "What happened to the cow?"

"Dad shot her."

"Oh! Why did your Dad shoot her?"

"Cause she kept getting out."

"Oh." I said.

I suspect that he was a little mad at her and thought he had a game load in for partridge, pheasants, pigeons or whatever, and thought he was going to teach her a lesson with a little bird shot. I can't imagine anyone trying to scare them with a slug! She must have been running to miss the other leg, as the entrance wound was on the inside of the leg. I couldn't sew it, as there was nothing there to sew. It was gone, completely gone. My alternative was to treat it like an open wound.

I got the yellow spray can out, the only nitrofurazone spray which is now illegal, and no longer made, and overlaid it with a screw worm bomb to prevent it from getting maggots from the flies. That was a lindane, chlordane product that in no longer made either (also illegal today). I put her on long term penicillin injectable, which we put everything on back in those days. I left him a note explaining what I did and what he should do as follow-up. In my note, I specifically used the word injury. I did not say bullet wound. I did not want to go there.

A signed check was on the bulk tank which I filled out and left a copy of the receipt. I said goodbye to Dennis, and left.

I wondered two things later. What happened to that cow, if she ever healed up I never heard and never asked. And, secondly, what happened to Dennis. Do you suppose he ever told his dad that he had told me, "Dad shot her."

That call would have been an excellent chance to show what amazing things Aloe Vera, CEG and my Super Wound Spray in today's organic world could have done to heal up that wound.

Please don't let Doc usher at early service anymore.

122

When In Rome, Do As The Romans

This article I'm not proud of, but I want everyone to know that no one is perfect and we do some dumb things in life no matter what our age.

I was raised in Southern Minnesota amongst a bunch of Germans, my ancestry on both sides is German. In my boyhood, the Germans around me in the 40s and 50s all smoked Camels, Chesterfields and Lucky Strikes. No filters, straight lung clogging cancer causing cigarettes. Little did our society know. We were fooled just like the GMO brainwashing we are in today.

Well, I get to Arcadia, Wisconsin, in June of 1967 and start my practice with a mixture of Scandinavians, Polish and Germans, and they all have a round can in their pocket, filled with Copenhagen. A high, high percentage of my rock solid, wonderful clients had their lip full of tobacco. One day, after many offerings from my good hearted farmers who would ask if I wanted a pinch, I decided okay, I'd try a pinch. I took a pinch. I'd never been a smoker, so when that nicotine and whatever else I absorbed hit me, I got dizzy. WOW! This stuff was way too strong for me!

I told someone of my experience and they said to start with a pack of Beechnut Chewing Tobacco and work up to Copenhagen. Only real hard core chewers could chew Copenhagen. So, that's what I did; and, yes, that's exactly what happened. I stayed with chewing Copenhagen for 13 years. My wife hated it. Looking back, I realize that she really must have loved me to put up with that awful habit.

I was a spitter. When it was in my lip, I was like Pavlov's dog. I started copiously salivating. I spit on car windows, pickup windows, the car behind me and on the sidewalk. You name it and I spit on it. Some people can put in a chew and go to sleep. Not me! I had a little oblong hole in my dash and two cans would fit in there just nice and tight. Why two? You never, ever want to run out. No sir. I had clients that knew of my extra stash and if they were out and weren't going to town, they would grab my spare full can and tell me to add it to their bill.

Well, one winter I developed a head cold/sinus infection and between every call I would put in a fresh chew, go about two miles and start gagging and almost throwing up. I stopped, pulled over and thought, 'How stupid am I?'

Here I was making things worse without even knowing it. I rolled down my window and sailed those cans just as far as I could. This was in March of 1984. That night I told my wife that I was done chewing, and I was. It was one of the toughest habits I ever had to break. In fact, my oldest boy who was in Junior High then said to me one day, "Why don't you start chewing again so you aren't so ornery." Unbeknownst to me, I was somewhat irritable and short during my withdrawal.

I reached into that little slot in the dash for a year. I thank my wife and family for putting up with me and helping me quit.

I'm still in Rome, but now I watch the Romans and don't do as they do!

Child Nurse

One Saturday, my wife, kids and I were invited to a good old time Polish wedding.

One of the joys of life is to crawl into your selected eco system and enjoy living it. Well, let me tell you, the Polish culture knows how to put on a wedding! I always enjoyed watching my older, bad kneed, worn out hip, farmers do the hop polka with their lovely wives. How those two could glide across a dance floor in perfect harmony – it was like they were floating on air. It was an artistic expression of two spirits, just beautiful. I would later see this man in his barn limping, waddling, bent over and think this was the man I saw doing the hop polka so smoothly! Unbelievable!

Okay, back to Saturday. I had some calls in the a.m. so I hustled to get them done so we would have ample time for the wedding, meal and the dance. I got done fairly early so had about an hour to spare when I got home.

I decided I would hand weed some little trees I had planted close to the yard. I had marked them with a steel electric fence post with a little sharp metal triangle on them, as I didn't want to use sprays anymore to kill the grass. I got down on my hands and knees and start pulling the grass, fast as I can go. About the fifth tree had the steel flat triangle in the grass, by the tree, where I had put it in upside down. I reached in with my right hand and tore a triangle flap of skin, about 1¾ to 2 inches, off from the top of my hand. It was a flap, like a tent flap, laid back behind my knuckles going from left to right. It was sort of neat, as I didn't cut any veins or tendons, and you could see them so nice. Sort of a lesson in anatomy! It didn't hurt much and there wasn't a whole lot of blood. The first thing I thought was, "Darn, I'm going to miss Kabunainski's wedding."

In my truck, I carried a little double edged half-circle suture needle for pig ruptures and small stuff.

Aha! I'll just throw a few stitches in that and then go to the wedding. I had it all figured out! I rinsed it off, put some fine catgut in my needle and I started stitching. The flap didn't hurt as I had cut all the nerves, but the intact part did hurt a bit. I get the first stitch in and had no idea how to knot it. Pretty hard with only one hand. My wife ran a convenience store in town and had gone there to do the banking, and I really didn't want her involved in this operation as I just suspected she wouldn't condone my course of action.

Our fifth child was home, a very capable young lady, strong and focused. I go over to the house and holler for her. She comes out on the deck and wants to know what I want. Very fatherly, I say, "Can you come over here by my pickup, sweetheart?" When she gets to me, I ask, "Do you love your good old Dad?" She rolled her eyes and says, "Yes."

She was our expert eye roller – picked it up at an early age and perfected it. I then asked if she could tie a knot, all this time I had my right hand behind my back. Another "yes." I explained the situation to her and said I'd stitch and she could tie. Her comment, "Bet I'm the only girl in my class stitching up her dad today. Give me your hand." I put in six more stitches and she did six knots perfectly. I got the Aloe Vera out, wrapped that hand up like a cow foot, and thanked my daughter. Everything was fine until mom came home.

I tried to explain that I had saved probably $900 in medical bills and we would have missed the wedding to boot. She didn't quite see it the same way I did.

My hand healed up without a scar, I missed no work, saved a lot of trips to the doctor's office and didn't miss my clients doing the hop polka.

Our daughter is now an acupuncturist and master herbalist. She is excellent at Eastern Alternative Medicine and can roll her eyes like an expert if it's warranted. I can't help but believe that I helped make her what she is today.

The Phone

My vehicle has a phone – it's the heartbeat of my livelihood come snow, sun or rain;
I'm convinced my phone has a mind tucked away in it, with a twisted little brain.

I was convinced this super device, invented before I was born,
Would be the epitome of communication, when I hooked it to my horn.

When I'm driving down the road, twenty miles between calls, my mind a blank,
Not one client goes to the phone, to give my number a crank.

But get 50 stalls away, with both sleeves wrapped in rotten afterbirth,
It'll ring so loud and so long, it just shakes the earth.

I was pulling a big calf, sweating, straining, my foot anchored on a rock;
The farm wife comes running into the barn and says "It's for you Doc."

I'm on a dead end road, and stop to empty my bladder;
You guessed it, that stupid phone will ring, clang and clatter.
Now, a stop sign approach, you're car number five, waiting so neat;
The look they give you when my horn blasts, makes me sink in my seat.

Just stop at a Quick Trip or a Fast Food for a sandwich and juice;
They all wonder whose little kid, is really cutting loose.

I'm changing oil, on my back, I've got ten minutes extra between clients;
Sure enough, the phone starts shaking the vehicle, like it's filled with a couple of giants.

The most fun of all, is visiting while the client makes out the check on the hood;
I'm always ready, but the poor farmer, his reflexes are amazingly good.

So, I'll keep on running through barns just to hear the receiver go "click;"
Cause when I get in and drive away, I know it won't even let out a tick!

124

She is experiencing a viral erosion of her intestinal epithelial mucosa,
resulting in a proliferative enteritis giving her an afebrile emaciating condition
resulting in a guarded prognosis. In other words, she's gonna die from the shits, Clyde.

Dr. Kabuna is on call at the present time. If you have a touch tone phone and have digestive
problems, press 1. If you have a reproductive problem or a calving, press 2. If you have a
young stock problem, press 3. If you have a delinquent bill or need some foot trimming,
hang up and call another veterinarian.

I'm chelating her. She's a little high on iron.

Today

Today I treated a Milk Fever on old Bess,
Picked up a foot and found a sole abscess.

Next was a physical, for severe indigestion;
I made a major ration suggestion.

Today I treated mastitis in the vein;
Farmer said "It's dry, we need rain."

Next was an Amish farm, looked at a horse;
Nice gentle creature, they use them, of course.

Today I treated a sow for MMA;
So sick, all she could do was lay.

Next, was a slow milker, open a teat;
That's a job where you watch the feet.

Today I treated a uterus that produced a calf;
If she doesn't improve, production will be half.

Next was a heifer, so weak and in pain;
She caught pneumonia, from a cold rain.

Today I treated a bull totally off feed;
He needs to get better, there are heifers to breed.

Next was a pregnancy check, to find the day;
To see if she bred in April or May.

Today I treated a cow for severe hardware;
She ate some metal, who knows where.

Next was a cold can of Mountain Dew;
I've got to watch my nutrition, too!

Today I treated a sheep with foot rot;
Cleaned it up and gave it a shot.

Next was a cow with a leg to inspect;
I left them some medicine to daily inject.

Today I treated a case of pinkeye;
That bacteria was transmitted by the fly.

Next I went home, supper was fed;
Crawled up the stairs, and went to bed.

126

Just what we need, Clyde, $11 milk and now Latino cows coming north.
or
This dairy industry will be pretty much Latino if this keeps up long.
or
The Latinos work harder for less. Do you suppose these cows will milk more on less.

Better call your banker. Looks like your herd is liquidating.

No! No! That's not what I said.
I said you should sell enough of your cattle to cut your debt in half.

What's New, Doc?

In the middle 1980s I'm heading to my next call, when my wife calls to tell me that the Kabuna Brothers have a cow down with Milk Fever and she rolled down a little incline through a fence and she is in the neighbors' hay field, flat out, almost dead, hardly breathing.

Now, these two fellas hardly ever got too excited about much of anything. When they would greet me upon arrival, they would always have a pat question – "What's new Doc?"

They traded cattle, milked a herd and had lots of livestock in every building. They were nice fellas, who had not only considerable wealth, but also a wealth of information! The older brother got around and was a walking newspaper and was always on the hunt for a new piece of gossip or news. Consequently, the first question was always "What's new, Doc?"

So, when my wife called and said the Kabuna Brothers were really excited, she knew this cow was serious, so I dropped everything and headed their direction. My wife called me on my bag phone. (That was pre-cell phone – a little tele-phone company had put some towers up and had some vets and contractors and a few businesses using bag phones. They worked pretty good unless you were stuck in some hole up against a hill. Bag phones were the first mobile phone marketed and I embraced them at that time as I was on the technological edge!)

I knew about where this cow was as I knew their farm, so I wheeled into the neighbors yard, cut along the edge of his hayfield and headed to the back side of Kabuna's farm. I spotted the brothers in the hay field with the prone cow, and yes, she was about on her last breath from Milk Fever.

Every once in a while, you'll hit one that runs out of calcium fast and her heart stops. Besides, the fact that the brothers like to hold onto their money pretty tight, she probably should have been treated the night before.

Now, this is all taking place on a Monday morning. On one of my earlier calls that morning, when I asked one of my clients if there was anything new (you see, I kind of like this opening as well) I learned that another good old 60 year old, red neck down on his luck kind of guy, had had a heart attack and had tipped over dead early Sunday evening. I jumped out of the pickup, ran to the cow, hit my IV and had calcium hooked up and was watching her heart intently. I didn't give a physical, get a temp or even say "Hello," as she was hardly breathing. After about 3 minutes, her heart rate starts to get stronger, she's beginning to stablize and I lean back, give a sigh and the older brother says "What's new Doc?"

I look up and say "Well, I heard Clyde died last night." (Knowing that I had a juicy bit that they hadn't heard yet, because really, Clyde was almost still warm. Usually, no matter what I told them they were already up on it).

They looked at each other with a shocked, wide-eyed look and the older brother said "Shit!" This stunned me. I had never heard either of these old gentle hearted brothers say one swear word – ever – before. Shocked I asked "What did I say wrong?"

The older brother looked at me with a sullen face and said "We just loaned him a hundred bucks on Saturday."

That new vet sure has a following with the cats.

Strides

When you answer the phone, you know who it is by
   how they talk;
If you see them on the street, you can tell who it is by
   their walk.

I have one client with a long, distinct stride;
He carries himself with dignity and pride.

A big fella, named Roy, stands six feet eight;
He wanders around without a distinct gait.

One old fella, who's kind of a bragger,
Just slides his feet along, I call him a dragger.

You see bouncy little guys, that walk with a hop;
The size thirteens, you hear every step with a plop.

One distinct leaner, who walks with a sway;
He lives on a ridge, he's got the wind all day.

A short little guy, who's always humming a song,
Has short little legs, that shuffle along.

How about the hitcher, long torsoed Mr. Black;
He's always pulling his pants up, they're way too
   low in back.

The next farmer has his toes way out when his feet
   he does plant;
Looks like he's always trying to step on an ant.

There's pigeon-toed feet, the toes they do bang;
Looks like a prisoner, tied to a chain gang.

The old guy with a limp, he's got a bad hip;
Always has a big wad of tobacco in his lip.

The crooked legged one, who's knee joint is shot,
Had better have surgery, or his whole leg will go to
   pot.

I see my little boy, his pants hang so funny and low;
I look in the mirror, and what do you know.

"Yes!" says Grandma, "He walks just like you."
Now I know what I'm like from the back view.

Age

Now I've hit fifty and feel spry and look pretty fair;
I'll admit my beard has some white and my
   sideburns have a few grey hair.

I had an old friend stop in just to visit one day;
My kids asked, when he left, "Who was that Old
   Codger anyway?"

I was stunned, my life flashed before me like a parade;
Why that Old Codger is only older than me by one
   decade.

It seems like only yesterday I was playing football
   and Buddy Holly was rolling and rocking;
Where do these boys think I got my good advice on
   tackling and blocking?

I'm not vain, I leave my grey hair, no pulling out
   with a tweezer;
The only category older than a Codger is an Old
   Geezer.

I asked my boys one day, to define what they
   seriously mean;
After we had passed an old man in his driving
   machine.

Well, a Codger is old, grey, but still somewhat in
   touch;
A Geezer is wrinkled and weathered and doesn't
   know much.

Now just a minute fellas, they may not know
   computers and rock bands;
Their back may be stiff, and they may have callouses
   on their hands.

They are the ones that helped build this great nation
   out of the dust;
They are honest, and genuine, and loaded with trust.

I gave them a lecture on faith, honesty, freedom and
   the flag we fly;
And heard them walk away whispering "Dad's sure
   an old fashioned guy."

So I will just wait as time will teach wisdom, let
   experience have its flair;
I will wait until they are fifty, and watch them get
   grey hair.

School

A rule at our house – Down by the road, don't make
  the bus wait;
Sure different now, I rode my bike two miles, back in
  1948.

Now the closets are full with extra clothes on the
  walls;
My mother ordered me two new pair of bib overalls.

A salad deli, pizza, nachos, specialty food to the last
  detail;
I took a couple of sandwiches, an apple, and one
  treat, in a pail.

For Phy Ed we played 'Kick the Can' or checked our
  traps on a gopher set;
These days kids swim in a pool or use a gym with a
  basketball net.

Showers and lavatories and fountains, your thirst it
  will quench;
The 1940s facility, was two-holer with an awesome
  stench!

Girls have volleyball, boys practice blocking, and on
  Friday night they play;
I went home after school, milked the cows and threw
  down the hay.

The neighborhood parents brought wood, some even
  donated coal;
Now we have an oil truck, up to the school it does
  roll.

The erasers were cleaned, water was pumped, the
  students divided the chores;
Now the janitors come early, and a night shift locks
  up the doors.

The homework, the testing to get State Aid;
I've got to keep working to make sure the taxes are
  paid.

I hope we're progressing, and that we are bettering
  body and mental collections;
Because we sure have made it complex, with Mill
  Rates, Referendums and Special Elections.

When I graduated, I was pretty naive, but ready to
  tackle the world with zest;
Now they have seen the world and are being
  recruited for the best.

Some complain, and say our educational system is
  really a mess;
Be positive, this generation is much better educated,
  we're making progress.

Michele

No way. I won't have any part of it.
Last time you invited the locusts, they stayed seven years.

Cats, Cats, Cats!

In about every 100th small town in America, there is some old widow or spinster lady that lives in a big old house, who loves cats. We had one in our town.

The first sign of too many cats was the neighbors could see cats sitting in the upstairs windows and the slow destruction of the curtains by the cats. After many months of that, and a few more breeding cycles, there were more cats in the windows not only upstairs, but downstairs and, by now, the curtains were gone. There appeared white cats, yellow cats, black cats and multi colored cats. Cats, cats, cats.

The next major sign as the months flew by, was a very strong distinct order emanating from the cat lady when she went to the grocery store in her not too clean outfit, to buy 50# bags of cat food. When she left the store they would get the room spray out and spray the store.

Finally the neighbors began to complain to the city authorities about her. So, our Chief of Police went to pay her a visit, and couldn't believe his eyes. The place was literally filled with cats and absolutely filthy. Everything was coated with greasy cat dung. Everything. The stench was unbelievable.

He called the county sheriff. Our sheriff was an older common-sense retired farmer who I had done veterinary work for when he was still farming. They could see that this needed immediate attention. She had committed no crime, as she was feeding them, but our sheriff talked her into going up to the hospital. I don't know how he did it!

Since our sheriff knew me and I had been running the dog pound for our city, I got the call for help. The Humane Society (our county

didn't have one at the time this happened so they responded from Eau Claire about 50 miles north of us) was called in and they rushed to the scene with one cat carrier that could hold two cats at the most.

What good they thought that would do I have no clue.

The sight that met my eyes upon entering that house is something that I will never forget. I had never seen so many cats in one area in my life! They had fleas, some had skin problems and some had watery eyes. I could sense that the Humane Society had no clue what to do and were as overwhelmed as the rest of us. Plus, these cats stunk really bad!

I made an executive decision that we would save 9-10 really nice ones (if we could find that many) and the rest would be put down. I got no flack on my decision as no one else had any clue as to how to proceed. I noticed that the sheriff

132

and police chief were putting Vicks Vapor Rub under their noses to fight the odor.

The first dozen or so were caught by being gentle and slow. After that, they became more like their primordial ancestors and hissed and clawed and ran. The entire police force was called in. Heavy leather gloves were put on and fishing dip nets were used on some to net them. I used a drug, xylazine, that is an IM tranquilizer that takes a while. I positioned myself by the kitchen table and as they held them down, I gave them a shot in the hind leg. They were then put into a big dog kennel to go to sleep. They were then placed in my pickup truck.

This entire project started at about 4 p.m. and when we had that last cat caught it was 10:30 p.m. We knew some had gotten away, which we planned on live trapping later.

My problem was this pickup load of cats was only tranquilized, not put down. I knew I had a lot of cats, as my truck drove like I had a ton pallet in back. I hustled home, drove into my garage and got the Pentobarbital out and injected them to stop their heart. I finished up shortly after midnight.

Now, this was a horrible job. I have always liked cats. In fact, we always had house cats and even had a clinic cat. These cats, however, were diseased, wild, malnourished and some were even deformed. There is no way they could have been saved and found homes for.

So, now, I was faced with a new problem. I was the only veterinarian, in probably the whole United States, who had a heap of dead felines in the back of my pickup. What could I do with them?

The decision came to me during my nearly sleepless night. I got up early the next morning (a Saturday) and about 4 miles down the road from me is a fellow in the earth moving business. I talked him into doing a little job for me even though it was Saturday morning. I told him he needed to come over with his rubber tired John Deere back hoe.

At that time, I had a 10 acre CRP field next to the house. I drove my pickup out into it, behind a heavy bunch of golden rods and unloaded my felines. I counted them as I unloaded them – 243 cats. Yes, 243 cats. At 7½ pounds average, that's 1822.5 pounds. That's almost a ton – bigger than a big dairy cow. No wonder my half ton pickup was swaying.

When my back hoe buddy drove in the yard, I led him up to the golden rods with my 4-wheeler. When he came around the corner and saw this pile of cats, his jaw dropped.

"Doc, where did you get all these cats?" He walked around the pile twice shaking his head, saying "Damn, I never in my life. Never."

I told him to just dig a deep hole and said I wanted the last cat at least 4 feet down and to send me a bill.

I went into the house and took another shower as the smell from that whole job was really hard to get rid of. I smelled cats for days after.

I had three live traps set, as there was an old shed out back and over four weeks time I caught 11 more cats. Those I buried myself.

The house had to be torn down as the odor would have stayed there forever.

Old Blue

I once doctored a Roan, with a swollen up tongue;
She had foam and saliva invading her lung.

I gave her two antibiotics, right in the vein;
We put her on pasture, she walked down the lane.

Three days went by, the owner said "What can we do?
"She can't eat or drink, she can't even moo."

I tied up her head, and looked at his wife,
I had never cut off a tongue with a knife.

It was purple and leathery and getting rotten;
I cut out her tongue, it was soon forgotten.

The took some calf feed, and made up some soup;
They put it in a pan, and set it by the coop.

She slurped and gurgled and drank her fill;
And went and laid down on an alfalfa hill.

She stayed in the herd for many years;
When I'd walk in the barn, her eyes filled with tears.

She's a dandy old cow, they named her Old Blue;
Now when she bellers, she says "Malou."

Jump, Doc, jump!

I think you are dehorning a little deep, Clyde.

Well, Tonto, now that we have wiped out the last milking mammoths,
what are you going to do with your separators?

I'm warning you. Please don't ask my husband again what a large animal vet does all day.

What a wonderful girl Junior is dating. I guess she'll eat anything.

Let's come back tomorrow. I can't stomach this fresh stuff.

Mistaken Identity

Back in the 1970s, when there were four of us operating out of a clinic, we had a 2-way radio to communicate with between each vehicle and the office. This was before electronics were born.

I finished a call and called in to the office. The office girl listed off the new calls, and one of them was really close to me, so I said that I'd get Clyde. He had a retained afterbirth.

Now, it so happened that we had two clients with the same first name and their last name was very close to the same, except two double letters in the middle were different. But, when you pronounced their names, they sounded a lot alike.

The two-ways were a little scratchy, sometimes hard to hear, so I pulled into the wrong yard.

This second client didn't call us regularly as we were second on his list of vet choices, but his regular vet was quite overworked and lots of times we would fill in. I thought this was the case.

I go into the barn to get my usual pail of water, whip into the stalls, he's milking, and I say "Hi Clyde, you got a cow to clean."

He looked up and said, "Well, yes. She's down on the left. I just milked her a minute ago.

I go down and go to work on the cow and she was really tough, way too early. I wondered why he called us so early as she was no way ready to be cleaned. I did the best I could and said "Clyde, we better come back in three days and retreat her so she doesn't get sick."

'OK," he said, "I'll pay you when you come back. Good day."

"See, ya," I replied, "we'll be back," and I'm out the door.

Next day, Clyde #2 calls in and asks "Why didn't you stop in and clean my cow yesterday?" He lives on the opposite side of town and kept the cow in all day and we were a no-show.

Then it hit me. Clyde #1 probably didn't call. I went back on the retreat at Clyde #1 and quietly asked, "Hey, what time did you call us?"

He looked at me and said, "We didn't.

"At the time you came, I thought my wife must have called you. When I got in for breakfast, I told her that Doc Dettloff was here. She told me that she never called you as it was a few days early to call anyone, as she had just freshened.

"We had no idea how you knew."

I explained my mistake, then adjusted the bill, and we had a good chuckle.

Light or dark this evening, Gentlemen?

Did you notice my new manure stacker when you drove in for the surgery?

No, I wouldn't call it pasture bred. I think it must be what they call pen mating.

What's In Your Trunk Today?

In 1979, when OPEC raised the price of a barrel of oil, the bottom fell out of the huge gas guzzling cars, and you could pick one up pretty cheap, as no one wanted one. The press had a party on gas guzzlers!

Well, I bought a 3 year old Mercury Marquis with 25,000 miles on it from a private party for $2700.00 cash. It had every whistle, bell and electric gizmo on it you could imagine. Surprisingly, it wasn't too bad on mileage, about the same as a pickup in the late 70s. My wife drove it for a number of years, trouble free. I change my own oil and when it got to 100,000 miles I decided to wear it out as a veterinary vehicle. It had a huge trunk and a big back seat so it was adequate for my needs.

I had a nice client who raised heifer calves and did some cash cropping. One day I drove in to do some work and here are 7 buffalo in the pen, 6 cows and 1 nice bull.

I remarked, "They are extremely wild you know. It takes a really good fence to hold them in." Then I half jokingly asked, "What are you going to do with your male buffalo head when you no longer need him?"

He laughed and said, "Oh, I don't know. That'll be a long, long time. The robe and head will be in big demand!"

About four months go by and I return to his farm. I ask how the buffalo are doing and he goes, "Not good! My bull disappeared – no tracks in the snow, nobody has seen him. He just vanished!"

Spring comes and I get a phone call. "You still want that buffalo head?"

"You bet! But weren't you going

Since they invented the Clovis Point, they don't care much about dairying, do they?

to be using him for years?" I asked.

"We found him when it warmed up. We could smell something in the back of the hay shed. (Which was part of the open shed where the buffalo were kept.) He crawled in between those big round bales, about 3 bales deep, and couldn't go ahead or back up. Got stuck and died."

They removed the bales and there he was – dead. They dragged him out. He didn't smell the best, but I figured it would have a beautiful skull. At that time, I had probably about 50 animal skulls – warthogs, beavers, chinchillas – I had a lot. Sort of the veterinary hobby, boiling out and peroxiding skulls!

So I put on two sleeves, cut the head off, got a feed sack out of the barn and threw it in the Mercury Marquis' trunk! Big trunk, just made for buffalo heads!

Two calls later, the Watkins feed salesman was on a farm as I drove in. He was a friendly fella. He came over while I was getting my grip and pail out of the back of the car and asked what was in the sack. I casually replied, "Oh, just a buffalo head."

"A what?"

I opened the top of the sack and said, "A buffalo head. Want to see it?"

The odor hadn't been bad until I opened the sack. Then, WOW!

The odor hit him and he apparently had a weak stomach. Wasn't pretty!

About a month later, after my buffalo head was all cleaned up and every bit as beautiful as I knew it was going to be, I drive into another farm client's farm and here is the same friendly, weak-stomached Watkins man just coming out of the barn.

He doesn't seem quite as friendly as he hurries to his pickup and says, on the run, "Please don't open that trunk until I'm out of here. I don't even want to know!"

Well, Doc, my problem is this. Ever since this mad cow thing, everytime I talk about culling someone for slaughter, they flip into this Downer Syndrome and I don't know who to sell.

Some Things Never Change

I do vet work for an ornery, grizzly old duffer,
Who every year, has one little black heifer suffer.

She's bred too young, the pelvis is tiny and round;
He should wait and breed them, when they are larger
    by one hundred pound.

He'll watch them labor, strain and groan;
About midnight, he decides to go to the phone.

"Get out here, Doc, bring your scalpel and tweezer;
Looks like it's time for another Caesar."

"Yep, Melvin, my old friend;
"No possible way it'll come out the back end."

"I knew it, Doc, I knew you would cut her;
I've already got you a pail of warm water."

An hour later, we have a live bull calf;
Laying in the corner, in the clean straw and chaff.

"I'll sew up the peritoneum, muscle and skin;
"Let her grow a little, and breed her again.

"Now, Melvin, next year don't breed 'em so quick."
"I know, Doc, this is a costly little trick.

"I'll guarantee it, I'll let them grow a little bigger;
You can count on that, for sure, by jigger!"

A year later the phone rings at twelve thirty and a
    half;
"Get out here quick, I've got a little heifer trying to
    calf."

Waves

On vet calls I drive my circles that never end;
I get many a wave from many a friend.

I like the big right hand waver, using his whole arm;
He's a real friendly fella, usually born on a farm.

I get a kick out of the waver that says "Hi" as he
    opens his mouth;
He must know I can't hear him, whether I'm going
    east, west or south.

The guy with both hands on the wheel gives the eight
    fingered job;
His thumb firmly on the wheel or the knob.

There's the one that never sees you as you come
    down the road;
His mind is occupied with a heavy load.

I always wave at him as we meet and pass;
I see his head snap as I look in my rear view glass.

The effortless, nonchalant head nod makes me blue;
I just know what he's thinking, "Oh, it's only you."

Then there's the flagger who waves his flat hand up
    and down;
I don't know if he's waving, or wants me to stop and
    turn around.

Another one is the wiper, his arm back and forth like
    a pendulum on a clock;
I look in the rear view mirror, and he's still waving
    as I see his head rock.

In the summer, when it's hot, you're wiping sweat
    with a sponge;
It breaks me up to see the one that out the window
    his head and arm lunge.

Then there is the one who owes me quite a bit of
    money on a delinquent fee;
I wave extra hard at them, as they pretend not to see.

In spite of who I meet, I'll just keep giving a big
    wave;
And intend to continue it, until I'm in my grave.

141

Hurry, Wendell, he's gaining on us.

Naw, this ain't no hardware. Looks like lightning to me.

CARTER

Oops!

There once was a vet
Named Dr. Bershet.

On a farm owned by Dutter,
He slipped and fell in the gutter.

He fell on his veterinary grip,
Gave himself 5cc's of Testosterone in the hip.

They just had their tenth little Bershet;
The hormone is still working yet!

Dead Cow Down!

One warm summer day the phone rang at noon.

"This cow had some mastitis this morning, Doc, so I kept her in to strip her out," Clyde tells me. "But she went down hill fast. She's really sick, Doc, can you come soon, as I want to go back out and cut some more hay."

I buzz over east of me, where there is a whole community of Norwegians and Swedes.

This client was a little bit older than me. He and his wife had an empty nest and they seemed to have a solid relationship and both had a great sense of humor. Well, his barn was a little older, so over the years he had lost all of his metal stall dividers as they rusted off. As they rusted, he cut them off at ground level and did without them.

After examining his cow, I found Clyde was absolutely correct, she was one sick puppy. Her heart rate was racing – 100 plus beats per minute. Scary heart beat. Her one quarter was as hard as an oak post and nothing came out of the teat, it was rock hard.

When I find one this sick, I look them in the eye and study them, and she had the look of death. It takes quite a few years to develop this skill – but it gets to the point where you are seldom wrong. When I see this, I tell my client what I think, that is that I'll treat her but she has a really good chance of not making it.

Now my standard treatment was to IV them with an antibiotic and put Oxytocin in it to empty the udder out. Then Dipyrone to lower temperature and an antihistamine for tissue damage, both IM. (This was before Banamine.) I would then mix up saline solution, 250cc, and spike it with Gentocin and run that in the quarter. (A lot of Gentocin went into cows before they could test for it.) So, I tied her up and ran the IV antibiotic and Oxytocin into the jugular. She did let down a trifle of milk. I told Clyde to strip that quarter out clean while I went out to my truck to mix up the saline and Gentocin.

I get out to my truck and can't locate the saline. I heard a yell and figured the cow must have kicked at him. All was fine when I'd left. Clyde had his head in her flank, above the udder, and was stooped over stripping her teat. I found the saline, and heard another louder, more desperate yell. I still wasn't too concerned – figured she didn't like him stripping on her sore teat. I get my prep mixed up and head back into the barn. I will always remember the sight that met me upon entering the barn. The cow was no longer standing, but laying down, dead, on her left side

with all four feet out, not moving a muscle, on top of Clyde. She had died, standing up, and with no stall dividers, she crushed Clyde under her as she fell over. Clyde's face was red. He couldn't push her off him and she was on his chest.

The poor guy couldn't breathe. His cows averaged 1500 pounds and his herd was primarily black Holsteins. Fortunately, I had a lot of upper body strength, so I grabbed her dead, limp tail and gave a big pull. As I pulled, Clyde would try and slide back out from under her. We gained about an inch each time. After about a dozen or more pulls, Clyde was free. Exhausted and out of breath, but free. He sat on the gutter edge with his feet in the recently cleaned gutter. He is gaining his composure back and his color is much better.

So, I, to add a little humor to the situation said, "Boy, did you have me in a tough spot."

He looked up and said, "You in a tough spot! I'm the guy she was laying on."

"No," I said. "She was laying on your checkbook."

You see, Clyde carried his checkbook in his Osh-Kosh coveralls in the chest pocket, and paid every call with a check.

He laughed and said, "I guess I'm lucky then that she was on my checkbook, or you might have left me there."

I left, didn't charge him and chalked it up to another life experience.

Well not so. Clyde always drank a little beer in a tavern or two in Blair, and when he hit town Friday night, did he ever have a story to tell. It was all over town that Dettloff said if she wouldn't have been on his checkbook he would have just driven away. Every client over by Blair heard the story and a common comment when I drove into a yard was, "Lost any cows by Blair lately, Doc?"

Boy, did that story get around – all in good fun, though.

He has passed on now, but every time I go by his old farm, I vividly see that big black cow with all four feet out, laying on Clyde.

Memories, memories.

Squaw Root

One of the most impressive tinctures I discovered early in my sustainable world was Squaw Root (also called Blue Cohosh or Caulophyllum). I started tincturing the root of this plant for reproductive uses. The earliest known use of this plant was by Native Americans who used the root to assist in the contractions of childbirth. That's how it got its name of Squaw Root. The dried root was chewed.

The veterinary use that it is employed for is in the removal of retained afterbirth. It can be put into an Aloe Vera liquid uterine infusion or given sublingual (under the tongue). In uterine infections post calving, it can be given vaginally for 5-6 days, about 3cc. It opens the cervix and gives the uterus tone to expel the exudates.

In 30 years of practice, I had never successfully expelled a mummified fetus from the uterus. I was able to remove them with caulophyllum. I did four in a row. Actually, it takes about a week to 10 days. One was expelled and three were brought up into the vaginal area, half way through the cervix.

Those I had to reach into with a glove and pull out the rest of the way. I would then infuse them with my Aloe mixture, but could not get any of them to breed back.

I have a younger colleague who I told this procedure to. He also had success in removing mummys this way. He, however, had one that bred back and conceived.

I could never get synthesized drug company hormones to remove a mummy but the natural products captured in a tincture would.

Lightning

There is no diagnosis that requires more common sense than when you encounter what may be a lightning strike. All insurance companies require a postmortem signed by a veterinarian before they pay a lightning claim. Nearly all farmers carry lightning insurance because it is not uncommon.

When I graduated, I had received about a two hour lecture on it by an instructor who had never seen a case. On post-mortem we were told the veins in the legs are full of blood. You'll see burnt hair, they will very likely be on a hill or under a tree. Yes, sometimes that is true, but I've seen some very odd things over the years.

I happened to be driving on calls one day and encountered a rain storm that was a gulley washer. I pulled over onto the shoulder with my wipers on high and still couldn't see much. I was sitting there waiting for the rain to subside, listening to my new 8-track with Johnny Cash singing "Ring of Fire" when one of the four big cottonwood trees about 40 yards ahead in the ditch gets struck by a jolt of lightning. My pick-up literally jumped. I could not believe it. It went "ZAP" with an instantaneous fire ball, a huge fire ball traversing down the tree in an instant. I was shaking after I saw it. I could see that anything close to that tree would be killed.

A few weeks later, I go by that tree and it looked quite normal. It had releafed and was standing there, not dead, not shattered. I was perplexed until I read later in an old veterinary book (lost wisdom) that there is a difference in hardwood trees versus softwood trees. The hardwood trees, (such as oak, hickory, locust), are much more dense and heavy and when the transpiring liquids in the xylem and phloem of the tree get struck there is so much heat that the liquid turns to steam and uncontrolled steam in an enclosed space expands and blows. Now a softwood tree, (box-elder, poplar, willow, cottonwood, soft maple) are not near as dense and there is more room for the steam to expand. You might have an upper limb blown off but you won't see them explode. I've seen an old tree that had big splinters, 20 ft. long blown 30-40 yards away. It's also not uncommon to see roots blown out of the ground.

How is lightning generated? In a normal scenario, the ground on the planet always has a negative charge. Clouds have a negative charge on their bottom side and the top of the clouds have a positive charge. That is why in the early stages of a storm, as you get a building of an electrical charge in the clouds, you will see horizontal lightning going from cloud to cloud, never having any ground strikes. When the charge builds up more, the clouds will reverse their charge giving the bottom a plus charge and the top of the clouds a negative charge. Then you will have cloud to ground lightning.

If you have ever been high in the mountains out West during a cloud build-up around a mountain peak because of warm air rising, and you get in an area where your hair stands up and you can feel it, get out of there quick as that is an area where lightning may strike.

The most I've ever seen killed from a lightning strike was 31 cows all gathered between a silo with a steel roof and a concrete re-rodded feeding area and a metal bunk with a metal roof over it. Thirty-one dead, boom, in one spot.

I had 16 Angus beef cows all under one big lone oak tree on a little rise in the pasture. I've seen animals along a barb wire fence dead, where the lightning struck on top of a hill, hit a tree and the bolt of electricity went down for about 80 yards through some low thorny brush where it burnt about an 18 in. circle through the brush and killed two heifers grazing next to a fence.

Usually the animals are killed instantly without any struggle. They are dead before they hit the ground.

I've had everything happen in my many years. I had one old, slow pay, client that hauled me way back in his pasture next to a spring where I posted an old cow that was full of hardware, with terrible infection. I posted her and said, "Sorry, Clyde, this is not a lightning. I'm calling a spade a spade and she died of hardware and had to have been really sick lately."

He looked at me and said, "If you call her a lightning, Doc, I'll pay up my vet bill."

I said, "Clyde, no, I can't do that."

He quit calling me for about four years until one day out of the blue, he called and I was his veterinarian until he died. We never brought up the subject of that lightning call, and life went on.

The Amish self insure with their super funds they have for fire and Acts of God. Twice I posted dead cows on Amish farms with the Bishop and the Elders looking over my shoulder. That's when I would verbalize my thinking process on what happened. It was actually a rewarding experience.

I once got called to a farm where a farmer was standing in his barn watching his cows come to the barn for evening milking in a light rain shower. Lightning struck a clay knob right next to the cow path coming down the hill and instantly killed three cows. A fourth one rolled down the hill and laid there dazed for awhile and got up and came to the barn. She kicked and kicked with her hind legs so bad, she was dangerous. He couldn't milk her. She also had a wild look in her eyes and would jump to any sharp sound. I saw the three dead ones, gave him a lightning slip. I said watch that other one and let's see if she settles down. Two days later he called and said the kicking and antics got worse. She wouldn't eat and then that morning she broke her left hind leg when she hit the stall divider. She had gone completely crazy. I saw her that day and put her out of her misery and gave him a lightning slip for her also. The insurance company became unglued, but I held firm on a phone conversation.

Insurance companies knew that they got worked over on the whole lightning claim issue. I had two old cattle dealers that had cattle all over in rented pastures in the summer. They would make a tour about every 3 weeks to look them over. Most of these were sale barn animals which now and then will contain a lunger. A lunger is an animal that has lost the battle with pneumonia and is slowly going to check out.

These brother cattle jockeys called their insurance agent and wanted a claim. He had been more than generous to them in the past and had issued claims without a veterinarian postmortem. He suspected he was being taken, so he sent me. I picked up one of the brothers and we drove 20 miles to the back pasture. We crawl up a hill and here lays a pile of bleached bones. No skin, no organs. The buzzards and coyotes had done a good job. I picked up a clean white lumbar vertebrae and tossed it on my dash. I politely explained to my good client that I could not issue a slip because I couldn't do a postmortem. He cleverly said, "Well, can you prove that it wasn't struck by lightning?" I said, "No, but in good conscience, I can't put down in writing that it was struck by lightning."

He could see I wasn't going to budge, so we small talked all the way back to his farm. About 10 days later, I see the insurance agent on a farm and he says, "Hey Doc, what happened with Clyde's lightning animal?"

I reached over on my dash and handed him this snow white, non-smelly super clean lumbar vertebrae and said, "What do you think?"

He just smiled and said, "I thought so, Doc. Thanks, thanks a lot."

As a person, I have gained a lot of respect for lightning. Colorado and Florida lead the nation in lightning deaths in humans. Respect it.

Doc's really tired out. He's on the wrong end again.

Must be something real serious. Doc's giving her a complete physical.

Houdini

Once a horse trainer bought this beautiful horse.
His intent was to train him to jump, of course.

His name was different: called Houdini, he was first-
  rate.
They couldn't keep him in the pasture; he'd unlock
  the gate.

In a pen he went, a rope they did wrap;
Next morning he's out, he opened the snap.

They double tied him with nylon in a little stall;
He was out eating grass before nightfall.

Into a horse trailer he went, locked with a clamp;
He opened that in an hour, the bedding was damp.

"I know what we'll do," said the trainer one day.
"We'll use a padlock, and throw the key away."

Into the metal cage built with bars and welding
Went the horse named Houdini, the beautiful
  gelding.

Next morning the trainer looked out and alas!
That smart young fella was out eating grass.

"That's it! We'll leave him out. He's real tame.
Now I know why they chose Houdini as his name."

Art

I know a young farmer by the name of Art.
He feeds his cows silage with an electric cart.

One day, in a hurry to make hay in the heat,
He took a corner too short and ran over a teat.

The cow kicked and bellowed, "For goodness sake!
Next time get a cart with a decent brake!"

Trees

I pulled over and stopped just to let my eyes gaze,
Over the beauty of the trees in these mid-October
  days.

The show of rich deep colors is like their curtain call,
Because we know how bleak the winter is, that will
  follow fall.

The red oak take on its coat, a deep scarlet hue;
They're scattered on the hillside, usually quite a few.

The ash, now they turn yellow, with an edge of
  brown;
The poplar, they come early, with bright yellow all
  around.

The walnuts come real early, their leaves already
  departed,
The squirrels are hauling walnuts, their winter
  storage started.

White oaks can vary, they are the browns, even rust.
There are three deer digging for acorns in the dust.

The globular hickory, the little brown nut, you know
  it well,
They are a squirrel's delight as they tear off the
  shell.

A tear in my eye, I see the pathetic dead elm trees.
Such a majestic thing, too bad we have Dutch elm
  disease.

It's a sight to see; a forest filled with ferns,
  blackberries and sumac.
A beautiful home for the birds, deer and creatures
  with fur on their back.

We better wait two more days, Buzzy. It says ten days before slaughter here.

Oh, goodie. This one's a bull. I get the left one.

The Nail

There once was a German who had him a horse;
He knew all about the blame thing, of course.

This horse came up lame, real lame, yes, you bet;
He went to the phone and called him a vet.

Now, I had seen lots of cows and one horse last fall,
But, I was the benefactor of this man's call.

"You're an old vet, long time at this game;
Can you tell me why this creature is lame?"

"Let's start at the bottom and work to the top.
If we can find the source, his lameness will stop."

With the owner's knowledge of lameness bubbling
   forth,
This horse cost him a mint, he bought it up north.

I proceeded to the leg and picked it up nice,
Took out my hoof knife and pared on it twice.

When all of a sudden I heard it go click;
I looked at the blade and it had a nick.

Let's examine this hoof that's lame on the trail.
Embedded in the frog was a rusty old nail.

These books you've all read, the knowledge you read
Is all for naught, there's a nail in this steed.

I fixed him all up, and looking real cool.
I had forgotten half his knowledge since I was in
   school.

Away I drove with my grip and my pail;
Thank goodness it was just a rusty old nail.

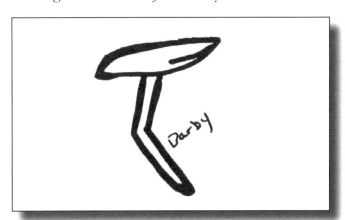

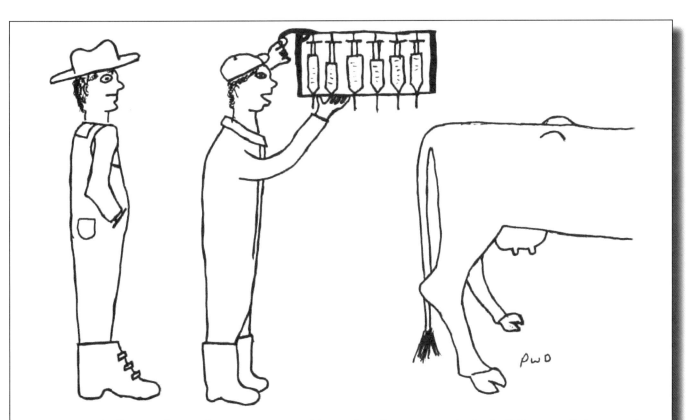

Naw, I quit diagnosing a few years ago. I just give them one shot of all the medicine made
with this handy little syringe I made and most of them are back on feed in a couple of days.

Let's name him Orkin.

Must be a cheese surplus again. Look at the size of that bait.

Tractors

When I was a boy, we had a tractor, John Deere
   Model B;
One reverse gear and forward was one, two or three.

A simple hand clutch, to put it in gear;
A rounded metal seat, that just fit your rear.

Two mechanical brakes, was all we did need.
When it was in road gear, Man! was that speed.

Hydraulics didn't exist, you did it manual, young
   fella;
Yet we were still modern, we had a green umbrella.

No radio back then – now they are a must;
We just hummed us a tune, as we drove in the dust.

Protection from the elements, there was none for sure;
You always rolled your collar up, when you spread
   the manure.

We cultivated corn, two rows at a time;
Planted clover and brome grass, no alfalfa, never
   used lime.

Hopped up into a big new green monster, tires taller
   than my van;
Thought I would quick move it, see how it ran;

Turned a key and hit a switch to get it going;
The air conditioner blasted my neck like it was
   snowing.

Pressed another switch, turned another key;
Now I have a rock and roll tape blaring at me.

It can't be this hard, don't start raving and ranting;
On came a monitor that told me to stop planting.

I'll just leave it sit here, driving around it is no chore;
Now, if I could just get out of this air-tight, self-
   locking door!

I guess there's no choke, petcocks or primer;
I'm beginning to see why my kids think I'm an old-
   timer.

Did She Have A Hiefer Or A Bull?

It was about 1969 and I was entering my third year of veterinary practice. I had two very fine classmates that were practicing with me – one was a partner and one was an employee and later to be a partner. We had built a new, modern dispensing clinic downtown and drove red pickups with veterinary boxes. Things were rolling along just fine in veterinary medicine in Wisconsin.

I inherited some real old time, conservative, unique, back in the hollar type people when I bought the house and practice lock stock and barrel in 1967 from an older veterinarian. One of these clients was a little wisp of a women, with snow white hair, that lived up over a hill and down into a little wooded cubby hole in the next valley. She was an older widow, a sweet lady that had a strong personality.

Her son was probably 50 years old and crawled out of the hollar about every two weeks or so to get some supplies. I doubt that he ever made it to Winona, Minnesota, (about 27 miles away). A couple of his more unique characteristics were: he didn't like to look at you, only quick little glances and he wouldn't talk to you only gestures and grunts. It was a game to come up with a question that he would have to answer and not use a grunt or point. A good question I used was "How old is this cow?" He said "6." I thought I won that round! Next time I asked, he won as he held up four fingers, telling me she was four years old. Next time, I won again, as I asked "Did she have a heifer or bull?" He said "Bull" – another round I won!

He would come into the clinic to get a box of mastitis tubes or whatever, and it was always written on a piece of paper. The office girl never heard him talk.

Well, one morning, the old widow woman called in with a Retained Placenta. The call went on my list as I was closest to the area of their farm. I got there probably around 8 a.m. and he had finished milking. When I arrived he was feeding a couple of calves milk on the other end of the barn, with his back to me. I had come in the south end of the barn and the animal to be worked on was a smaller heifer, the very first one on the left, with her afterbirth hanging out after a difficult calving. The milk house was on the north end of the barn, as was the house.

I set my pail of warm water and disinfectant down, as we carried warm water in our trucks, opened my medicine bag and commenced to start to remove the placenta. The heifer moved forward in the stanchion. She was small to begin with and was difficult to reach, so I stepped across the gutter, up about two inches onto the stall where she stood. The cow to my right was a fairly big cow, so here I am hidden away on the end of the barn, not saying a word as he won't talk anyway, peeling the placenta off her uterine cotyledons.

My quiet friend finishes feeding the calves, puts the pails down and saunters down toward me. He gets two cows away and stops, reaches up where there is a big nail pounded into the 8 x 8 oak post upon which sits a roll of toilet paper that I had never noticed before, turns his back to me, straddles the gutter about five feet from me, unhooks his Oshkosh-by-Gosh coveralls, lowers them along with his boxer shorts, squats and starts a huge, biologically normal, defecation process – in full view.

I discovered he was quite proficient at producing noise from both ends as he had a loud, beautiful grunting noise coming from his vocal cords and noisy exhaust from the other end!

This heifer was quite easy to clean and I'm done, but he isn't. This folks, was a major movement, so I faked it, holding the unhooked placenta hidden away, waiting for the completion of his process. After three copious wipes, he stands up, pulls up everything, buckles up his Oshkosh-by-Gosh, steps across the gutter back onto the walk facing east, turns his head south and spots my pail.

Meanwhile, I'm busily holding this placenta. He looks up at me, I look at him, his eyes get as big as coffee cup saucers. This was not an amorous look I got, more like pure panic.

He turns north and heads up the walk, out the door, hoping never to be seen by me again. I didn't have the heart to tell him he forgot to put the toilet paper back! I replaced the paper on the nail and left. I thought it best to leave and send the bill rather than chase after the money.

The next morning when we were dividing up the calls, my partner asked me "Well, did you get Clyde to talk?"

"No," I said.

"Sit down, I've got a story you won't believe!"

Your turn, Clyde.

Who's Discing?

In my practice, I had a nice long valley with 14 dairy farmers in it. They were all my clients. Of course, neighbors watched neighbors closely to keep an eye on things.

Early one spring day, I turned up the valley as I had two veterinary calls there that morning. The very first farm, which at the time was a very good set up where things got done, was a sandy farm that was the first to dry out in the spring. As I approached the farm, the owner pulls out of his driveway, goes across the road to his lightest sandiest piece, and starts discing.

I wave, he waves, and I continue on. Just around the corner, over a little hill on the second farm was my first call. As I get in the barn and start to give my patient a physical, my client says, "What's new Doc?"

"Clyde's discing." I answered.

His head snaps towards me and he says, "What?"

"Clyde's discing," I repeated.

Now, this farmer was a little gossipy and watched his neighbors like a hawk, so I knew this piece of news was putting him in orbit.

I finished my call, he paid me and I bid him farewell. I could tell

he really didn't want to do any visiting, so I headed on up the valley to the second call. It was a call that took about 1½ hours as there was a lot of fertility work and calf work. After I was done, I headed back down the valley. When I got to the farm where I had my first call, I look over and here is a 4020 John Deere with a disc, tearing up some heavy clay – way too wet to work up, getting ready to plant.

I chuckled and thought to myself, 'Next time I'll not say a word about Clyde so I can let that heavy clay dry out and not interfere with anyone's schedule.'

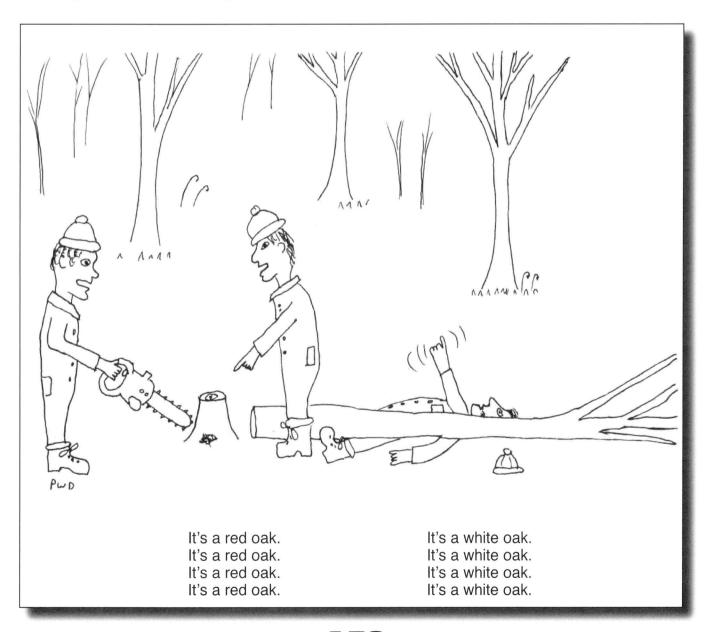

It's a red oak.　　　　It's a white oak.
It's a red oak.　　　　It's a white oak.
It's a red oak.　　　　It's a white oak.
It's a red oak.　　　　It's a white oak.

Absolutely Nothing

Let's start this poem – but I've nothing to write.
Went to bed early, and slept real good last night.

Woke up this morning, my mind clear, I'm not a
  crank!
Picked up this pen to convey a thought, my mind is a
  blank.

I'm driving – looking – there's a big stainless steel
  milk truck,
I try these lines, nothing rhymes, on that one I'm
  stuck.

There's a big blue heron standing in a clean little
  stream,
No let's see, empty thoughts, I start to daydream.

The trees are turning colors, it's really pretty out;
Can't get anything to click, I'm so frustrated I could
  shout!

A squirrel just took a chance in the middle of the
  road;
I would have squashed him, if I hadn't slowed.

This poem is sick – there's no philosophy, humor or
  satire;
So I will quit for today, lay the pen down and retire.

Let this be a lesson – if you are having a bad day;
Don't get frustrated, back off and walk away!

My Assistant Will Help You

After the OPEC oil scare in the 1970s, when Nixon lowered all the speed limits to 55, gas guzzling cars were like having the plague.

I bought a 1968 Mercury Marquis with every whistle and bell on it, for $2700, with only 25,000 miles on it. It was a battleship grey and nearly as big. It got like 13 miles per gallon, about the same as all the pickups with 300HP which were selling new for triple the price. We used it as a family car for years. Our friends called it a Ghetto Cruiser.

After about a 100,000 miles I decided to wear it out in practice. The back seat was huge and the trunk was even bigger and worked well for my medicines and equipment.

The only problem was that it was really heavy in the rear end so it had a little drag and I had to adjust the headlights lower as everyone blinked at me for having my headlights on high beam when they were on dim. I figured this was pretty cheap mileage. If you could drive a car for two cents a mile and change your own oil, a person could spend a few cents on gas. The car never cost me a cent in repairs until one day.

A client, way out on the east end of my practice, called with a cow with a retained afterbirth. As I was driving there, the road was closed, as they were replacing a bridge and they had a detour over a narrow little road, hardly used, that went up and over a steep hill. As I was climbing this hill, my car went Ka-Thunk! The engine stopped and the car started to roll back down the hill. I steered it into a hayfield, got out, and looked things over. My drive shaft was completely twisted off. I called my father-in-law to come pick me up. He told me he'd be there in a while as he had to finish something up quick. I called the wrecker and gave him a verbal map of where he could pick up my car and what was wrong with it.

About that time, an old local legend pulls up. He was way into his autumn years and was a gruff spoken man of wisdom. He was the local dehorner for a lot of farmers for miles around and was a real character. He asked what the problem was. I told him. He asked where I was going. I told him. He said to grab my stuff and he'd take me there. Now, this character had a heart of gold but, because he had accumulated quite a bit of wealth and had an odd sense of humor, not everybody loved him. He had people who didn't understand his humor and people that were envious of him. I didn't know if the farmer we were going to was going to appreciate this guy's sense of humor or not.

As we drove up to the barn, the farmer and his son came out to check out who was coming in the pickup – they were expecting my Mercury. When they got to the pickup, my driver said boldly, "Hear you got a cow to clean."

The farmer looked somewhat bewildered and said, "Yes."

My driver said, "My assistant here will take care of it for you."

I grabbed my grip and went to the milk house for water. The farmer and I went into the barn and worked on the cow. After I was done, I wrote up the bill, and he wrote me a check. Meanwhile, the driver was visiting with the son. I got into the truck and away we drove with the farmer and his son standing in the yard with a puzzled look on their faces.

Stanchion Disaster

In the late fall of 1968, I was called to a small farm owned by a little conservative Polish farmer. He said he had pneumonia going through his herd.

He had a little, tight, small barn with no fan for ventilation. It had gotten quite cold with our first wintry blast, so his permanent pasture herd was housed in his warm, tight little barn. Then, as always, it warms up and they got a good barn pneumonia.

My little farmer client weighed about 120 pounds soaking wet, and when driving by I would see him taking short little swift steps. I had been in town for about 18 months and this was my first call on his farm, so I knew I was under scrutiny.

Every cow was running a fever, coughing and breathing fast. I discussed my treatment protocol which was to give the most serious ones an IV in the jugular vein. This was to be about half of his sixteen cows. I did notice that as I was checking these cows, they had all quit eating and were totally wide-eyed, focused on my presence. I would imagine that I was the first stranger to be in that barn in about a year.

The lay out of the barn was such: he had two rows of eight cows, eight facing south and eight facing north. The alley between was about eight feet wide with tails facing tails over a fairly deep gutter. The stanchions were home sawed full 2 inch oak, probably constructed in the early 1920s. The stanchions were oak in a 'V' shape. A cross double member on top secured the eight to the west wall and a 6 x 6 oak timber on the east.

As I took my pail and treatment up front of the south eight, the first cow spooked and pulled back. Mind you, these were huge 1600 pound or bigger Brown Swiss. Then the domino effect with the second, third through the eighth cow, pulling back and jerking the entire stanchion complex completely off the wall, tipping the post over and causing the heavy home sawed mess to come crashing down on their necks, knocking half the cows down and the other half bellowing and pulling and pushing and thrashing every which way but loose!

My little Polish client opens the door and escapes with his life. I start carefully, to try to open the stanchions and untwist their heads one by one, without getting killed from their front feet. I told my little unhappy farmer to let them outside. I theorized that they needed fresh air.

After much labor, pinched fingers, kicked shins and bruised ego on my part, they were all free. My therapy changed as I decided we would not go IV but go with an IM shot and put some drugs in the feed.

I offered to help pay for the barn repair for his stanchions, which he quickly declined as I think he wanted me off his property. He sold his herd about two years later and by that time the necks had all straightened out and he was back to returning my wave when I went by his farm.

From then on, I always looked to see how good those stanchions were fastened to the wall before starting any treatment!

Now, let's have a little talk about production, butterfat, conformation and repeat breeders.

Cars

The car makers are attempting to associate their
    wares of tin,
With the animal kingdom to hide the loose bolts and
    workmanship of sin.

I just saw a Colt going down the road-a,
It's got an American label, but made by Toyota.

There goes a Bronco down the freeway in the fast
    lane.
I fail to see any resemblance to a horse with four
    feet and a mane.

Another analogy that takes imagination past my
    brain,
Is how they can call a neat little sports car a
    Mustang.

That isn't all Ford did to the equine set,
For years they made a Pinto in one color and aren't
    done yet.

Now, Dodge is no better, they became tough in one
    giant leap.
They are using the Ram from the wild, wooly sheep.

I see a cat with big fangs. Mercury sales were good,
When they placed a Cougar's profile on every car's
    hood.

Buick has a trained hawk on its ads every week.
The Eagle is a four wheeler, made by Jeep.

I've got some suggestions for the automakers'
    engineers;
Here are some designs that will please the eyes and
    the ears.

I'd take a big pickup, put horns on the hood, with a
    brass ring to pull.
Make it real bulky and mean looking and call it a
    Bull.

Then I would design a car. To fill gas the whole
    fender would lift up;
And name it a Dog or maybe a Pup.

How about a sweet little driving machine that I'd
    call a Dove;
Sell it only to couples courting, or ones that are in
    love.

Let's produce a big, fat one with the front like a pig
    snout.
Call it The Road Hog, we could shake our fists and
    at it shout.

It's probably best I transferred out of engineering
    school back in '63
And went on to receive a veterinary degree.

You must have emptied her out by now.

No Respect

If I were a cow I wouldn't kick with my feet,
Unless, of course, you slapped an ice-cold inflation
  on my teat.

My tail I would never attempt to slap in your face,
Unless, of course, you didn't leave that pitchfork in
  its place.

I would always get up if you had the courtesy to say
  'Please.'
But, I don't appreciate your hard old shoe on my
  knees.

I would gladly take my stall instead of playing that
  game with my head,
If you would keep it halfway clean, you forget that is
  my bed.

How about some of that silage you give us, all soggy
  and laced with mold;
That slimy old water cup, I would like it fresh and
  cold.

And how about that tavern stop, coming in late with
  half a jag;
Here we poor dears stand, dripping milk from an
  engorged bag.

Why don't we hit the gutters? Because we've a
  lesson to teach,
When you throw that scrumptious ground feed just
  out of reach.

Another thing about the manger – kids, dogs, vets –
  it's a major route;
Then you come slopping along, with who knows
  what on your boot.

Production's down, balance the ration, reduced
  intake – that's all we hear;
You just try to relax and produce more with hard
  rock music in your ear.

Another complaint of yours is those silent heats, you
  say;
About the time we have the urge, you're on the back
  forty cutting hay.

And secondly, you sold the bull; he was our good
  friend.
Now, you send us no pleasure, in a pipette that won't
  bend.

I guess you could say from the farmer we get no
  respect;
That's why we get ornery and stubborn, we have our
  dignity to protect.

How Things Got Named

You're sitting too close to the fire, Bon.

How Things Got Named

Look what Gelding got caught going through the brush.

How Things Got Named

Hey! What's Nute doing to that poor dog?

158

Mental Wanderings

My mind sometimes wanders as I'm tooling down
  the road.
It's sometimes a relief to have thoughts that don't
  give one a load.

Like, where does pain go when it disappears?
And, why do rabbits have such long ears?

Usually there's no answer, the questions are so odd,
But, sometimes my mind takes a real sharp nod.

Like the other day I came around a corner and saw
  a Paleo man
With a stick in some sand, he was drawing a plan.

He said to me, "You go up on yonder hill
And chase down the herd, we've a mammoth to kill."

Lo and behold, I gazed on the slope to the west,
And there was an entire herd nearing the crest.

So I grabbed my spear, with a stone head so sharp
  and clean,
And headed barefoot due west, right thru the ravine.

I surrounded the wooly creatures and drove them
  back down their same tracks,
To see my Paleo friend with others, spearing them in
  the thorax.

I was congratulated by my friends for a job done so fine,
And they offered me some choice small intestine.

When I saw them eat, it made me drop my jaw;
For they were eating that poor creature raw.

I took out my buck knife and cut out the choice loin,
Took a match, built a big fire, for them all to join.

I roasted some to a medium-well savor,
Gave them a bite and said, "Taste the flavor."

The fire scared them, as I had it a flamin'.
They instantly knew I must be their Shaman.

They had never seen metal that could cut so thin, not
  thick,
And fire controlled, and started with a stick.

Then the mobile phone rang and to reality I came;
I left my Paleo man, his mammoth roasting in the
  flame.

But, maybe eons ago he did make a kill
Just like I saw, just west of that hill.

So I'll just keep driving, my mind cleared of things
  that don't matter.
We'll see tomorrow if the next valley won't be better.

Old Nell

There once was a cow named Old Nellie.
She had a big, long tail and a large, round belly.

Hay she loved, the ground feed she would devour;
She could eat a forty-pound bale in half an hour.

She was milked twice a day, fifty pounds at a crack.
She had good feet and a nice straight back.

Always took the same stall, No. 44,
Next to the alley, by the silo room door.

Quiet, gentle cow, never did fuss or kick;
Owner was calm, never did use a stick.

Nellie and the farmer, they had a good rapport;
If she ate all her concentrate, he'd give her a little more.

Took her to the Fair in the town of Gibbon;
Best in her class, brought home a blue ribbon.

Had twin bulls, named Barney and Eugene.
One was gentle, the other one was mean.

Two of her heifers are in the barn, pretty as can be;
They came from A.I. and have quite a pedigree.

"Best cow in the herd," owner says with a smile;
"Hope she stays around yet quite a while.

"Such a good cow, she's never had Milk Fever
Because my ration is balanced by Charles LeFevre.

"She's a cow with genetics that's really gifted;
If I had more like her, my mortgage would be lifted.

"Every cow is different; it's not hard to tell.
I doubt if there will ever be another Old Nell."

No problem. Most cows with a little metal in them
respond quite well to a magnet. I'll be right over.

OK, OK girls, line up. Let's see which one of you had it.

Look at those dandy ribeyes, son.

Tough Ones

It was twenty below on the 30th of December;
LeRoy Smith had a prolapse, I distinctly remember.

The buttons were big, swollen and round;
Those on the bottom were froze to the ground.

The poor cow looked like it didn't matter;
I said, "LeRoy, get me ten gallons of hot water."

We bathed it up, and got it all free,
Not a speck of dirt could we see.

I replaced it, pilled her, and put in a stich;
We rolled her over, away from a ditch.

She jumped up and away she ran,
I picked up my towels and went to my van.

Sometimes when things really look rough,
You are blessed with an Old Boss that is pretty
tough!

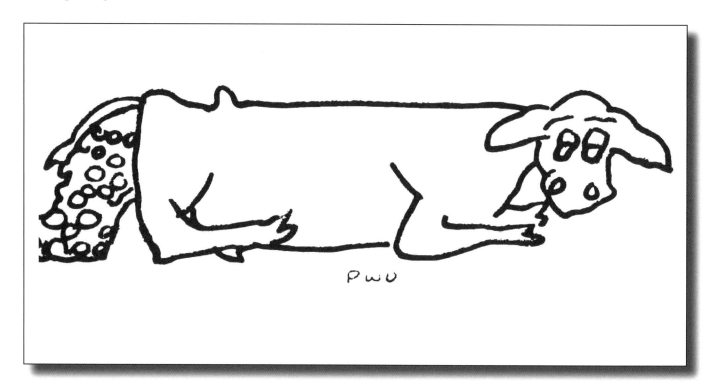

Guernsey

I checked an old Guernsey, thin as a rail;
She was owned by Louie and Dorothy McFail.

She was mean as a tomcat, ornery as sin;
Always got through their fences, she was never in.

"Give her a physical, Doc, she's not eating her
grain;
Maybe she's got something wrong with her brain."

I examined all systems, careful as could be; I said,
"Louie, this cow's older than me."
When you looked into her mouth not one tooth could
you see.

"Oh, yes, she's old, when she's born I forget."
"Well, Louie, you better take her to market."

So he loaded her up, to town he did go;
Didn't hope for much money, 'cause the market was
low.

The very next morning, she stood by the shed;
The only thing wrong was a wire scratch on her
head.

"Well," said Louie, "she got out and came home;
I guess we'll just let the old girl roam."

For five years more, that Guernsey lived, content as
could be;
Until one day, she died under an old oak tree.

The Professor

In veterinary school, we had a wise old professor from Germany who taught Obstetrics. He told us to never, ever amputate a cow's uterus as they will go into shock and die. I filed this in my memory bank.

I had been in practice for about a year when I got this call from an old, tall, quiet German. He had a very large farm in the end of a valley with a lot of woods and about 50 Herford beef cows to run out his autumn years. He had milked cows in his earlier years. He was the quiet frugal kind that had CD's in the bank. He found this tall, lanky Hereford heifer in the pasture with something hanging out, so I get the call.

I get there and get a pail of warm water and he walks to his H Farmall tractor and says, "Hop on."

I stand on the drawbar with my filled water pail and my medical grip, grab the back of the seat, put the pail between my legs, medical grip in my other hand and hang on for dear life.

We go all the way up the valley, crawl up a steep bumpy dug way, get on top and slide down into another valley to where about a dozen Hereford heifers are grazing. He tractors up to them and sure enough, one of them has a long, brown, cylindrical tube resembling a baseball bat sticking out of her.

I forgot to mention that I had my cowboy lariat over my shoulder as I was pretty sure I'd need that. I knew I'd only have one chance at roping this girl. As luck would have it, I roped her around both of her horns on my first throw. I snubbed her up to the front end of

his Farmall H really quickly.

She took off. When she hit the end of the rope, her back end spun around wiping her baseball bat rear end with a Mickey Mantle swing. I deduced that it was pretty securely attached.

Upon close examination, I see that it is what's left of a once prolapsed uterus. All the cotyledons (buttons) are gone and there is no blood. It's about 3-4 inches in diameter, covered with what looks like brown dried leather and it is a little bigger at one end. No lie, it looked like a Louisville Slugger, and it was almost as hard as hickory.

This prompted my first question. "How long has it been this way?" The reply was, as I expected, "Don't know."

I conservatively estimated 10

days to 2 weeks. I then explain to him that we normally would give her a spinal, wash it up, invert it like a sock, medicate her, sew her up and they usually heal fine. I explain to him this is not an option due to its condition.

I lay out two options. One would be to butcher her and turn her into hamburger or the second option was to amputate it. I prefaced this by saying that I was told in college that if you cut them off, they would go into shock and die. I told him I didn't medically see how this could happen as we don't have much tissue mass left to bother her losing it.

He said to amputate. I gave her a spinal just to deaden it and cut through the leather section on both sides to locate both middle uterine arteries. I then tied them off with some great suturing moves, went down about two inches and cut the uterus off. Surprisingly, there was very little bleeding. I then put a stitch in that I could invert, sewed her up and pushed it back into the vaginal area. I put one final stitch around her vulva. I had preserved the bladder and urethra. She humped up and passed her water.

I was on top of the world. It had never been done, but Dettloff, the surgical genius, had done it! I thought to myself, 'This should have been on film.'

I undid my quick release Honda, she walks away about 30 yards with her head down and stares at us. I put my equipment away, rinse out my pail with what little water is left. I get ready to hop back onto the draw bar and I notice she is starting to breathe fast, with little short breaths. Clyde sees this also. She breathes faster and faster, staggers about 10 feet and tips over dead. Stone dead.

I couldn't believe my eyes. Clyde could but I couldn't.

Onto the draw bar I went. Not a word was said, only this time Clyde's going faster than before. I think to punish me! Up the dug way, over the hill, down the dug way to my pickup, not a word. I'm too busy hanging on to his tractor seat. I sense that I may want to eat this call. He pulls up to my pickup, looks back without a word. I get off and say there will be no charge for today. He answers in one word, "Good."

As I drive away, I hear my professor saying, "Never amputate a uterus. They go into shock and die."

You sure can move a lot of air with that barn fan, Clyde.

Barn Doors

To get into every barn, there's a challenge when it
    comes to the door;
Some have only one entrance, some have two, some
    have more.

There're silo rooms, feed rooms, cow yards and a
    pen with a gate;
They all have some concoction; shutting them you
    have to negotiate.

The older the barn, the more ingenious the catch;
You see very few with a knob, spring and a latch.

The standard handy closure has got to be twine
    string;
It's either wound round a nail, tightly knotted, or
    strung through a ring.

Another one that's pretty common is the old bent nail
    trick;
You put your shoulder into the door, reach up and
    give it a flick.

The one least original, but it works, so don't it you
    knock,
Is the guy who simply takes his foot and slides away
    the cement block.

You see it occasionally, twine replaced it, and that's
    the wrapper bailing wire;
It's stronger than twine, quite scarce, but it would
    last through a fire!

The top barn door with one hinge loose, sort of a
    flopper;
That requires a pitchfork or two-by-four to close it
    and be a proper.

The sliding board that lines up just right, and slips
    in a crack;
Now that, my friend, takes some skill and requires a
    little knack.

The pole shed door that's moved in spring, the frost
    has played a game;
You pound and twist and then you pry, to close it into
    its frame.

I've seen nails, twine, wire and sliding boards, yes
    siree;
But never have I seen a barn that's locked with a
    knob and a key!

Doors

The American farmer has always been known for his ingenuity. This was displayed on a little hilly farm where one of my little Norwegian farmers scratched out a living with 15 cows and 12 sows. If any of my farmers was a hog whisperer, he was it.

He had an old dilapidated hog house where he farrowed his sows in pens. As the pen partitions slowly deteriorated over time, he would replace them with wooden doors – yes, wooden doors. Usually solid wooden doors from houses. Every farm has saved doors from when they remodel or just need new ones. Hinges pull out or door is no longer square, it gets saved as who knows when you'll need it? Farmers generally don't waste much. So, there's a farm auction. Out of the upstairs in the granary

comes grandpa's and dad's old doors. Four of them, all different sizes. Two of them have porcelain knobs and there are two on each door – inside and outside. Clyde saved all these porcelain knobs in a 5 gallon pail. Do you have any idea as to how many different porcelain knobs there are?

At least once a year, I would ask Clyde if he had any neat new knobs. "Oh Doc, you won't believe this brown and white one I just got." Did you know they made brown and white stripped porcelain door knobs? I didn't.

He had lots of white and lots of black, solid black. Then there's brown and it goes on from there.

One time I was called to his farm to examine a sow and surgically correct two ruptures. Upon getting out of my truck, I asked Clyde

what was new. He excitedly told me about the farm auction he was at the week before. He bought four doors with six porcelain knobs for a buck. Don't tell me there isn't excitement out in the country with a find like that!

He would talk to his sows. If one was sick, and you had to get into the pen (with 3 sides made up of doors cobbled together) he would get an ear of corn and hold it to her so she could eat it if she felt good enough, and you could give her a physical, check her udder and she wouldn't give you any trouble 'cause Clyde was right there telling her to behave because I was there to help her. He knew how many teats each sow had, how many pigs she had last litter, how many litters every one had had – every item about every sow was in his head. He truly was in sync with his pigs. The cows, however, were a different story. He was truly a pig man.

Before you got to his farm, Clyde would write down all his questions on the inside cover of his little match book. Clyde was a chain smoker, and had a match book in his pocket as a note pad. He could get 10 questions on one little match book inside cover.

He didn't live to be very old and I was one of his pall bearers. He was a very wise, unique and clever man loaded with common sense.

He had a spring coming out of his farm and of course, Native Americans camped and lived by springs. He had a pretty nice collection of Indian artifacts. I hunted artifacts also. When he would find a new piece, he would always show me. It makes sense that anyone who had an affinity for arrowheads would also have an appreciation for porcelain door knobs.

Those Clydes are now gone from the scene. He was the most gifted and profitable swine farmer I ever met. Clyde – the Hog Whisperer.

Can you imagine today, George Hormel and Company giving a contract to a man who rebuilt his pens out of old house doors and treasured the knobs. Funny thing about it all, is he had the healthiest and happiest hogs in western Wisconsin back in the 60s and 70s.

The factory farmed swine of today would have loved living and being taken care of by my Clyde in the hills. My, how society on the farm has changed. We've gone from ear corn and door knobs to computers and purrs virus.

God rest your soul Clyde, you had it all figured out.

Odors

One day a call came from Ralph Losinske;
He had a cow fresh a week with an RP.

When I walked in the barn, you know how it goes;
You can walk up behind her, by using your nose.

The odor, Wow! It made your nose sting;
The sloppy old tail I tied with a string.

While reaching in her, my arm to its root,
She strained twice and filled my left boot.

I washed with soap down to my feet,
Got in my car, decided to eat.

I stopped in a busy little roadside café;
In fifteen minutes I drove their business away.

The owner said, "Doc, you're a heck of a good guy,
But just for today, why don't you eat on the fly?"

So, I hopped in my car, down the road I did drag
Eating my burger and fries out of a bag.

Who stinks back there?

166

Son, you're old enough now. Put this in your arm pits each morning.

If You Got To Go, You Got To Go!

As I mentioned earlier, the high school boy that worked for us at the veterinary clinic was an excellent worker and sometimes rode with us as we enjoyed the company.

We'd usually arrange it so that it was a morning call or an afternoon call as all day could really get draggy for the person riding along. A half day was a long enough stint so that he, at some point in time, would have to empty his bladder. If someone was around the barn or yard, he would quietly disappear and reappear. Sometimes, if the farmer was in the field doing field work they would leave me a note so I'd know no one was going to appear, and in these cases our helper would just use the gutter in the barn to relieve himself. Many times I, as a joke, would pretend someone had appeared and say something like, "Well, hello Sue" or "Hi, Gladys, how are you today?" and watch him panic as he quit mid-stream and zipped up his pants. I would than laugh. After while, he became immune to my "Hellos."

One day, a neighboring veterinarian was on vacation and we were called to fill in at one of his bigger client's farms. It was only to work on one cow, which was quite common back then. So, I'm getting ready to work on this cow, and my helper went down further into the barn to relieve himself. I loudly say, "Hi, how are you today?"

Little did I know that the Mrs. was only about two feet from the door coming from the house and she answers, "Hello, Doc," and opens the door.

I was totally caught off guard, but not nearly as stunned as our helper was. The Mrs. totally knew what was going on as she watched him struggle with his zipper, but she graciously ignored the situation and held the cow's tail for me.

From that point on, I never pulled that stunt on him again, and he quit relieving himself in barns.

167

I used to have an old bottle of growth hormone around here.
Have any of you boys seen it lately?

My Country Boys

My boys have a fort down by a winding little stream.
They play there by the hour, in the humidity and the
   steam.

The inhabitants they know better than a book;
There are three types of minnows that live in the
   brook.

The Sucker's on the bottom, the Chub's in his spot;
The prettiest Minnow, by his gill has a red dot.

The Crayfish flipping away to hide under a stump,
You always hope when swimming, not to get bit on
   the rump!

That neat little bug that walks on top of the water;
It's fun to watch him zip around; where he goes it
   doesn't matter.

There's a hole in the bank, the boys can't reach the
   end.
A Muskrat lives there; he's their furry little friend.

The tracks in the sand, five toes so neat,
They belong to a Raccoon, with tiny little feet.

The dams that get built with sticks, mud and sand,
They fill up a little pool. They think it's just grand.

Then it gets filled up with tadpoles, from buckets and
   pails,
And for weeks they watch them grow legs out of their
   tails.

There's a plank bridge to get over to the other side.
But, usually they jump on the rocks, not breaking a
   stride.

A swing hangs out on a limb of a willow,
When they tumble off, they should have a big pillow.

When supper is ready up to the house they glide.
"You guys are absolutely filthy. Go clean up outside!"

Then we hear who fell in, the new dam that was made;
And Mom fixes the wounds with a kiss and a band-aid.

In the evening the sound of the frogs croaking away,
Makes me think the boys had another good day.

168

Guided By Spirits

In 1957, the day after Thanksgiving, my father's number was not up, and I'll tell you why.

I was a sophomore in high school. My older brother was in the Navy in Japan, so I was my father's right hand man on the farm. We had 18 milk cows and 8-10 sows. We sold the cream and the skim milk was fed to the feeder pigs in long wooden troughs. You've never seen excitement until you see 30 feeder pigs, 40 pounds each, drink skim milk in long wooden troughs.

Mother had 250 Leghorn laying hens. The egg truck came every Thursday. Mother had an old wooden ironwood stump with two nails and a hatchet. That's where our Sunday chicken dinners came from!

We were an average, little 120 acre, flat dirt farm in Southern Minnesota, in a church going German community. My great-great grandfather homesteaded in 1871 and I was born into a hard working, honest handshake society. This was the typical agricultural settlement of the time.

In the Spring of 57 it was cold and wet, and when corn was supposed to be planted our fields were mud. The crops were planted very, very late. We then had a late, late fall. The corn was slow in maturing and a frost, to kill the corn plant so it would ripen, was 2 weeks late. This left us with green corn stalks that were hard to harvest.

We had a one row pull type Case Corn Picker that we had behind a Model B 2-cylinder John Deere tractor, followed by a small gravity box for the ear corn. My job was to haul the gravity boxes up to the yard and unload them into the corn crib and after one round we would switch wagons.

Because of the picking conditions, the frozen, brittle corn stalk would hit the snapping rolls and break the stalk off near the ground and would get clogged up in a big bunch and not go through and shuck the ears off. My dad would get off and push them through these snapping rolls and then go a ways and repeat the process.

Now, snapping rolls are long cylinders set at a slant. They are about 5-6 feet long with longitudinal humps on them. They spin towards the middle. The one on the left turns clockwise, the one on the right counter-clockwise. So, when a corn stalk hits it, ZIP! it whips through. When it hits the big cob of corn, it separates it from the stalk and falls into an elevator and back into the wagon.

My dad had these snapping rolls set tight, which means they were only ½ inch apart. Really whipping at high RPMs. On the first three or four loads, my dad would be about 40% of the way back on the corn row. We had rows 40 acres long so it was nearly a load. The weather was absolutely miserable. A light snow, dark overcast day with the wind out of the northwest.

I had on my sheepskin coat. Back then we had no cabs on trac-

Woman, some days you really bug me.

tors. My tractor was a big old LA Case. About the fifth load, when I got back out to the field, my dad wasn't even done with the row going down. He was having big, big trouble getting these corn stalks to go through.

I don't know what possessed me – if I was cold or if I was told to, but I shut off my tractor and started to walk down to my dad. When I got there, he had just started back towards me. He nodded, I nodded and walked around behind the gravity box and jumped up to look into it to see how much corn was in it.

All of a sudden the B John Deere and picker went into a huge growl and my dad hollered. I don't even remember what he said, but I knew what had happened. I raced around and pulled the hand clutch and my dad's hand was in the snapping rolls. He calmly said to get the big crescent wrench out of the tool box, disengage the power take off and to see if I could turn it backwards so he could roll his hand out. I did as directed and couldn't budge him. I tried again – no luck. I played football all through high school and was well muscled, but I couldn't budge him.

We had a neighbor about a half mile away. My dad said I should go and get our neighbor real quick. I ran like a halfback, even though I was a slow lineman, to the tractor, unhooked the gravity box and was just coming to the field driveway when along comes our rural mail carrier. My dad had been on the school board with him for years. I stopped him and told him the situation. He said to hop in. He sailed down through the corn rows and we were there in no time.

He said I was to take the crescent like before and that he would take the belt pulley and to give it all we had. On the second try we rolled my dad's right hand out of the snapping rolls.

Now, my dad was a big framed man with large thick hands. When I saw his hand I thought there was no way they would ever put that back together again. The mail man said to have my mom call the Post Office to let them know he was taking my dad to Mayo Clinic in Rochester, Minn. We only lived 21 miles away. I can still remember my dad holding up his injured hand with his left hand at the wrist, as he gave me instructions. He told me to take the picker home and put it in

the machine shed where it always went and to be sure to drain the radiator on the "B" as it didn't have antifreeze in it. He was sure the neighbors would come and pick the corn so said I should build the extra snow fence crib by the hog feeding area where it always was and to tell Mother he'd be okay.

When I walked into the house, I swear my Mother knew. She knew something had happened. She asked what had happened and how bad was Dad's hand. I told her it was bad, but assured her that Dad was fine and didn't appear to be in much pain. I told her he was on his way to the Clinic, so she finished her kitchen work and left for Mayo Clinic. My younger sister and I were left in charge of the chores.

When I think back, my Dad had just gotten a new brown lined canvas jacket and he had it buttoned up tight around the neck. It wouldn't have taken long before he would have been pulled into those rolls up to his neck. My father had his right hand amputated. He was fitted with a prosthesis and did very well with it the rest of his life. I often wonder why I walked down to that picker on that cold snowy day. My dad's number was not up!

Half-Time

Now, I'm a large animal vet and I treat mainly cows.
I'm short on experience and expertise when it comes
*   to bow-wows.*

I treat a few simple things, but do a lot of 'referring.'
When it's a real sick dog I have no idea what
*   organisms are stirring!*

I'm lucky because most of my dog work comes on
*   Sunday afternoon, for some reason,*
Usually on into the fall when it's the pro-football
*   season.*

To be more specific, and this has some reason and
*   rhyme,*
The phone starts ringing when it's exactly half time.

I lose most of the work when I say, "Sure, I'll look at
*   Rover.*
Sounds serious to me, rush him right over!"

It's amazing how many recover, or get over their
*   cough*
About the time of the third quarter kick-off.

Now, I'm not facetious, nor do I hold a grudge.
But, I always miss the third quarter getting my own
*   popcorn and fudge!*

Vacation

Every summer we take our family to a cabin on a
   lake.
The adults go along to relax and watch their skin
   bake.

The day before you leave, it is always the same,
You're swamped; half the cows are sick and the
   other half lame.

You get up extra early to organize the kids and pack
   up the gear;
Then a neighbor stops in with a pig with a
   hematomed ear.

Finally, away you head, north, without a care.
Little Suzy is crying, she can't sleep without her
   bear.

The rustic old cabin is all squeaky and clean.
You hope the mosquitoes don't find the hole in the
   screen.

It's now Tuesday, you're relaxed and just had a thirst
   quencher;
Along comes a lady and sits down with her
   Doberman Pincer.

"Oh, you're a vet. I know you'll want to hear about
   Rover's left testicle."
She covers his medical history, every abscess and
   pustule.

"My vet is Dr. Earchop, he's the best in the state,
   with the best training.
But, would you look at this rash? It's lately been
   draining."

The free advice one gives doesn't have to be too
   intensive.
They never listen to you when it isn't expensive.

Lately when I travel, I go very low profile.
   After I've talked and visited quite awhile,

When they tell me their occupation, they're the best,
   yes, thanks!
I tell them I drive a truck that pumps septic tanks.

Spring

It's April in Wisconsin, we just had a nice rain;
It blew in from the west of the Minnesota plain.

Look at the hayfield with all the dandelions a-bud;
For a couple of days the barnyards will be mud.

There's one fella seeding oats, with alfalfa and
   brome;
One more round and he can head for home.

They all plant corn when oak leaves are big as a
   squirrel's ear;
The fields will be planted with a four row behind a
   John Deere.

Line fences will be checked, the wire pulled tight
To keep the heifers in the pasture during the night.

The stiff old cows, so lame and sore,
Will act like calves when they go out the door.

The birds all return, the orioles with their nest;
I really enjoy the barn swallows, by far the best.

The geese are going north up the Mississippi flyway;
I saw a pretty wood duck just the other day.

The pocket gophers are digging those mounds they
   build.
Oh, that poor rabbit on the road got killed.

The ringworm clears up from the spring sun;
To see the young stock frolic looks like fun.

On the edge of the woods stands a fawn in the first
   row;
Wisely watching from the thicket stands the mother
   doe.

The farmers, smiling, are busy hustling around.
Their attitudes are good, not a grumble do they
   sound.

I guess in the whole world there isn't a thing
That cheers everyone up better than Spring.

Let's take cover under this rock, it's been here forever.

Tonto, oh Tonto!

Time

Have you ever laid on your back on a clear summer
   night and looked up at the stars?
Now, I mean to look past our little niche, light years
   past our moon and Mars.

Do you have any idea of the magnitude of a light
   year?
No! We're too busy with trivial things, saying 'Oh,
   my!' 'Oh, dear!'

How about a billion stars? Now, that's a huge
   amount of celestial dust!
Infinity has no edges; it's not like Columbus sailing
   our earth's crust.

Time is so large we don't realize things will change.
Right where you're sitting now may someday be a
   mountain range.

Our planet may turn to ice or an inferno of molten
   heat;
It will be a celestial tombstone of time's relentless
   retreat.

Our mind's so small, our technology but a speck of
   knowledge;
I feel we're still in pre-school civilization; someone
   out there is in college.

I hope I can live long enough to shake the hand of
   some Extraterrestrial Soul,
But we're not yet advanced enough, I've not enough
   time to reach my goal.

So, I will leave this poem for some wise person eons
   later
To substantiate the truth, when we're checked on by
   some Invader.

In the meantime, I'll lie on my back looking up at
   Venus and Mars.
I'll just wave at whoever's coming to see if we're
   still using inefficient cars!

Trouble On The Trail

I just hate arm pits.

173

I don't know what Ugg De Laval thinks he'll accomplish with his
Centrifugal Rock Separator. Nobody will ever use it.

I don't want you going out with those boys from The Udder.
All they want to do is drink.

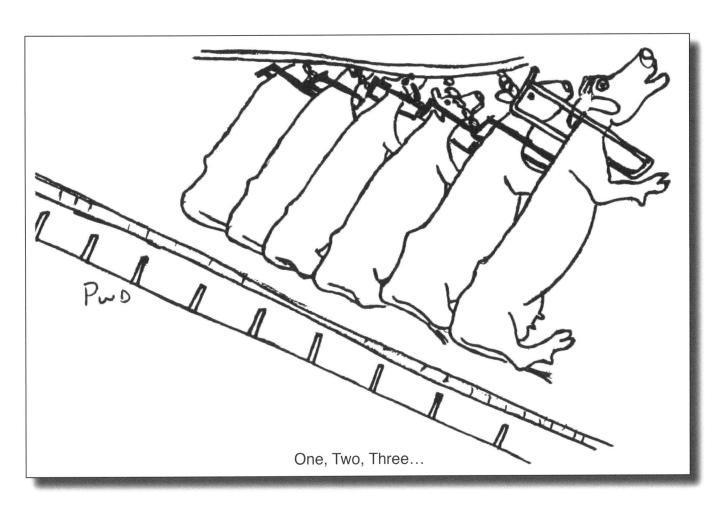

One, Two, Three…

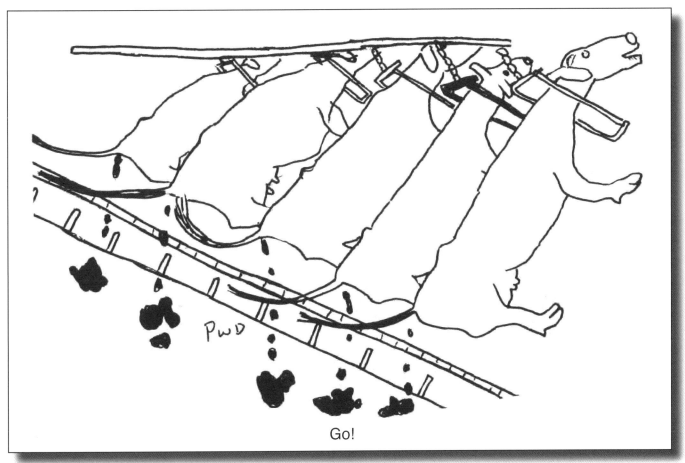

Go!

If

If I were to be a piece of old iron, some distant day,
I wouldn't be a gas engine, threshing machine or old
   Chevrolet.

I don't think there would be anything that would
   make me feel greater,
Than to be a heavy, old, cast-iron cream separator.

What kind, you ask? Well, it doesn't matter to me;
Just as long as I'm complete, including wrench and
   cream key.

I'd gladly be one of the Ankor Holths, or even a New
   Prima.
Just pour in your whole milk; I'll give you your
   cream-a.

Maybe a Melotte, with that funny bowl a-hanging;
Make sure I'm real level, so nothing starts banging.

I'd be a Sears or a Montgomery Ward, too,
Or even a Galloway, made down in Waterloo.

Tabletop or floor model, I don't care which,
But when I'm done, please don't throw me in a ditch.

My label could be brass, painted or a nice decal.
If you keep adding oil then you are a pal.

'Cause, you see, my insides are all very nice;
They are all made of brass and extremely precise.

How about a Delaval, with a really old patent date;
I'd rather not be a 17, something rare like an eight.

I could be a Diablo, something massive and brave;
Just turn the crank, that raw milk I crave.

A newer console, is Streamline, with tin on the side;
A beautiful place, for those spiders to hide.

If you are going to be an old three legged model, be
   careful of tipping;
I'd still give you rich cream, ready for whipping.

Let's call me a Canadian-made Renfrow, I could
   learn to say "A,"
Nothing would make me happier than to spin me all
   day.

When it's all over and I'm totally, totally rejected;
Don't sell me to a junker for scrap, let me be collected!

Barns

I sure see lots of barns, some white, most red;
Way in the back pen the calves have their bed.

The roof can be rounded, gabled or a high peak;
Covered with shingles or tin, so they don't leak.

The wood can be pine, ash, maple or white oak.
It's got to be strong so the floor joints don't get broke.

The older barns have character and class;
The new ones are so plain with concrete and glass.

Old ones have room for spiders and creatures abound;
The old horse barn has no cement; the floor is ground.

Old ones have neat boards with staples and a ring;
The new ones – there's no way you can tie up a thing.

Cows would hope that I wouldn't mention
The fact that they all have a wood or metal
   stanchion.

The manger is their table where they take their meals;
Some get fed with big forks, or out of a cart on wheels.

The bullpen is filled with too many calves, oh, my!
As most cows are bred artificially by A.I.

The silo room is damp, dark, dingy and cold;
A perfect place to grow fungus and mold.

The milk house is shiny, clean and protected,
'Cause periodically that gets super inspected.

Gone are the milk cans and separators to crank,
Replaced with a shiny refrigerated bulk tank.

New barns have silos and electronics that stray.
Old ones have home-sawed oak filled with timothy
   hay.

Progress, innovation – I guess it's all change;
Now it's NFO and Farm Bureau, no more Farmer's
   Grange.

Imports

I try to buy American; I was born in the USA.
I bought a Dodge mini-van and got foreign anyway.

'Cause when I opened the hood to check the motor
   and fan,
The motor said Mitsubishi, made in Japan.

You give a magnet so they'll eat the feed that's
   ground.
It's made overseas, but it's still black and round.

I like Abomasum Surgeries, I think they're a good
   deal;
The scissors I use are made of Swedish steel.

I order some Lepto vaccine from an Omaha plant;
The new owner is Arab; he talks in a chant.

I'm listening to my car radio, going to fix a pony;
Stuck in my dash, it's all made by Sony.

I buy new coveralls to keep my jeans clean of
   diarrhea;
I read the label and see they are made in Korea.

When I look in the field I see Deutz and Same.
John Deere is on strike, won't be settled today.

Here I'm turning a bull into a steer;
I check on the Burdizzo and that's not made here.

It's tough not to buy foreign, no matter what you do;
There's still nothing better than Red, White and Blue.

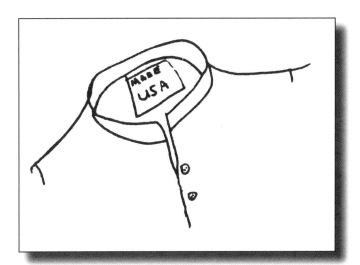

Nice soft stool.

UPS

An amazing company; it's well run, I guess,
Are those brown trucks that come from UPS.

When I stamp me an envelope, with tongue in cheek,
It may get there in three days, maybe a week!

Just tape up a box for UPS in zone three,
And it's delivered the very day that they guarantee.

I can be out in the boonies, mired in snow and muck,
And down the road comes that brown delivery truck!

Notice the drivers; they are all the same;
There's never a fat one, there's never one lame.

When one makes a mistake and sends the wrong bag,
You simply correct the mistake with a yellow call tag.

I've gotten boxes from California, Maine and North
*   Yoken;*
Hardly ever do you get one that is crushed or broken.

The substitute drivers are as good as the regular guys.
One was a blond Miss, with pretty blue eyes.

I send my blood samples, order my drugs and
*   penicillin;*
I know it will be delivered next day, with no mess
*   from spillin'.*

When I'm depressed, gloomy and down on my luck,
My spirits are lifted with a wave and a smile from
*   the UPS truck.*

So keep up the good work, delivering packages and
*   freight.*
It proves that our capitalistic system is still really
*   great!*

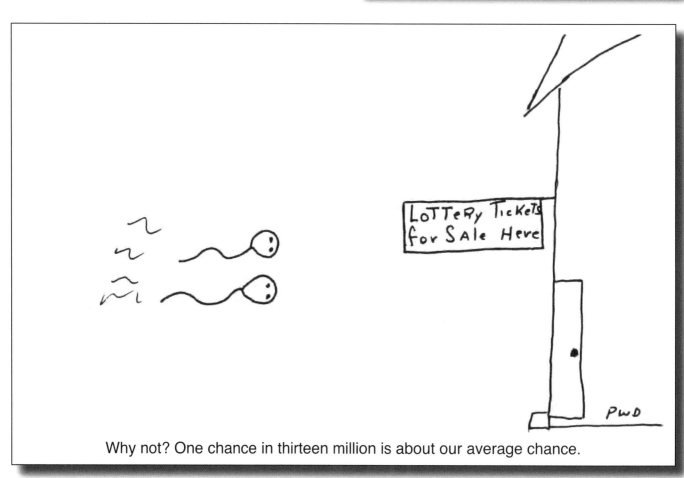

Why not? One chance in thirteen million is about our average chance.

Clyde

There once was a timid farmer named Clyde.
Whenever anyone drove in, he'd run in the barn to
   hide.

I treated cows in his barn, always from a note;
When I was done, my diagnosis I wrote.

One day I was stunned, he called on the phone;
A prolapse he had, I knew he was alone.

In the middle of the yard he stood, so excited;
He was so nervous, his fuse was ignited.

"Get me some warm water, we'll put in this womb."
For two pails he ran away like a zoom.

I washed it, replaced it, and sewed her up tight.
"Now you cut the stitch out about Wednesday night."

I looked around and Clyde did vacate,
Out the door, through a pen and slipped beside a gate.

So out with the pen, a note I did write;
It's been three years since he's been in my sight.

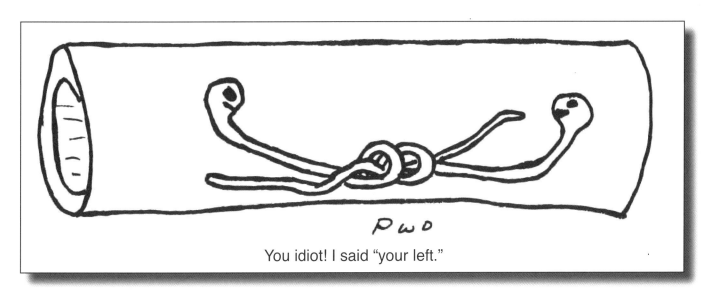

You idiot! I said "your left."

Medium, rare or well done?

179

Since they started having pig roasts every new moon, it sure is uncomfortable around here.

Well, there's only two of us left, and now you tell me you're gay.

The Predator

There are a few things I would like to explain today,
    on this earth.
One question I commonly get is, why do cows want
    to eat their afterbirth?

Let's remember thirty thousand years ago the bovine
    was evolving from the auroch,
They also had a predator watching them from behind
    every rock.

So, they would sneak off and deliver in some remote
    place
And then hide the evidence so no predator could
    trace.

They couldn't bury it, burn it, or set it on the curb,
    like garbage in town.
The only quick way to leave no trace was for them to
    eat it down.

To this day cows go to the corner of the woods to
    deliver,
Or down on some secluded bank close to the river.

Another question I hear almost every day,
Is why do we have flipped stomachs? We call them a
    D.A.

Well, way back the rumen was made for roughage,
    grass, legumes, even little twigs;
To get them to milk we push corn and soybeans; they
    are fed like pigs.

We've changed their diet, we're trying to
    nutritionally disgrace 'em.
Consequently they can't handle that much in their
    abomasum.

Now we measure butterfat, pounds of production,
    and protein levels for cheese.
Eons ago they had to run swift and produce enough
    milk for one calf to please.

Why do we have to have horns? Cattle should all be
    polled.
Well, that was their only means of defense as next to
    the glacier they strolled.

We've taken the bovine and changed her to fit our
    selfish needs.
We've forgotten she was once wild like us, back
    when we made huts of reeds.

Our perspective is twisted; the bovine does so much
    for us. We don't credit her.
And yes, we have forgotten that man is the ultimate
    predator.

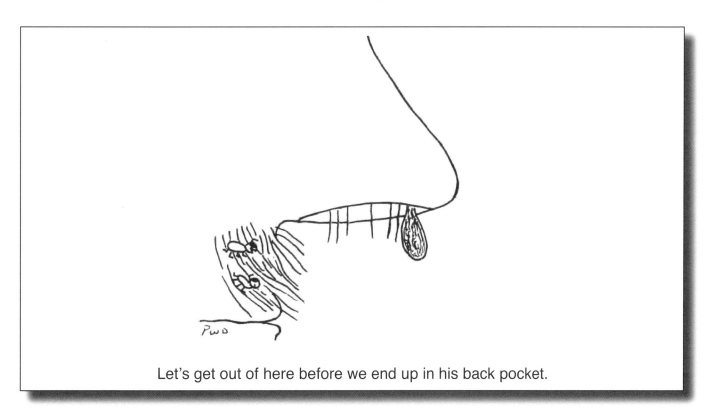

Let's get out of here before we end up in his back pocket.

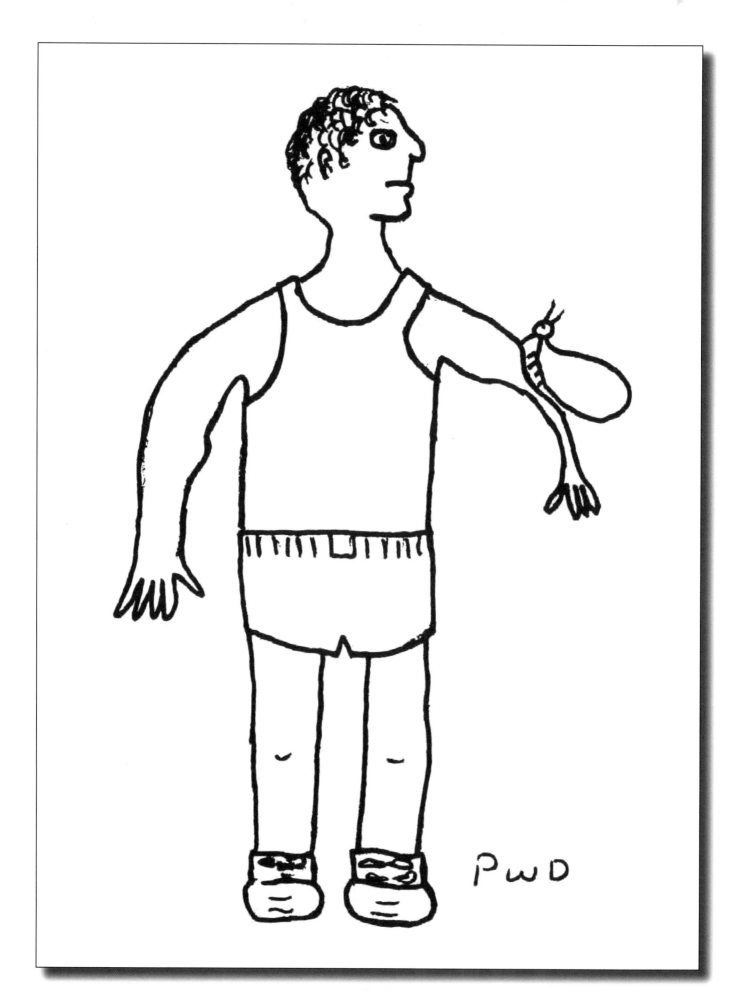

182

Society

I read of cultures past, I'm a nut on history.
Our society today, to me, is becoming a scary mystery.

The Classic Greeks became lazy and missed the omens.
Shortly thereafter they were run over by the Romans.

The Romans became Epicureans and missed
    history's pages;
Shortly thereafter, the civilized world was in the
    Dark Ages.

Our national debt exponentially climbing with our
    social spending,
I hope we can level it off, before it's the ending.

Some welfare folks think I'm somewhat of a jerk,
Because my laboring background makes me want to
    work.

They are a negative producer; a statement I recall:
The sum total of what we produce is divided by all.

Our throwaway economy, we've bigger landfills than
    golf courses;
I wonder how long before we're out of natural resources.

The weekends are for recreating, drink beer and
    watch TV.
Isn't anybody working, or is it only me?

Both spouses working – can't get by on one
    breadwinner.
Whatever happened to Grandma's good old-
    fashioned Sunday dinner?

The psychiatrists are busy; the social workers help
    the lonely grieve;
We've lost the family unit and didn't see it leave.

I guess as parents we are to blame, as we still call
    the shot;
We are all caught inside our society; we can't see it
    go to pot.

So step back, listen up, you patriotic souls; I've
    something for your ear;
Speak out, get back to logic and common sense,
    because I like it here.

Looks like Santa's been hitting the rum again this year. He just came down the hay chute.

What kind of clown do you think we have for our new hired hand?

What a nice herd of Van Goghs, Clyde.

Four Wheeling

I got a call at noon one summer day, after a few days of rain, from one of my regular clients telling me he had a milk fever across the road where his dairy cows were. I got there not too long after he called, only to find a note on the seat of his 4-wheeler along with a check. He had run into town for parts, and I was to take the 4-wheeler and treat his down cow, fill in the check and leave the bill on the bulk tank, and thanks.

Now, at that time, I did not own a 4-wheeler, but had ridden on them and had driven a few so it was no challenge – or so I thought.

I loaded up my equipment, motored across the road into his dry cow pasture. Most of his cows calved back in the far end in a steep little ravine in some trees. So I headed up a steep hill going in that direction. I get to a crest of a grassy knoll, hit the top of the hill, and start heading down to the cow. I didn't realize how steep that baby was or how slippery the rain had made it. I slid about 70 yards down, down, down. I slam into the narrow ravine at about a 40 degree angle. My pail of water spilled as the cow watched Evil Knieval come to treat her.

I treated the cow, and she was one of the 70% that dramatically responded, got up and headed home.

Now, the Honda 4-wheeler had a fairly big box elder tree in the ravine next to it. I tried pulling it around, thinking I could goose it around the tree. Ravines on steep side hills are always clay, really slippery clay, when it's wet. I managed, after a little persistence, to get clay and mud all over the 4-wheeler – really got it mired in good. I ended up walking back down the path behind the cow toward his barnyard. I filled out his check, put the bill on the bulk tank and left.

Next call there, he complimented my on my 4-wheeling skills. Thankfully he thought it was quite funny. Nice to have clients with a sense of humor.

No. No. This is the lame one.

Cow Tales

Me? Oh, I once kicked the hired man in the crotch. Man, did he scream.

I think Junior has picked up worms from eating all that fresh stuff.

Cow Tales

So the boss tells the vet, "Don't worry about her kicking.
We've been working with that teat for a week and she hasn't lifted a foot yet."

This Actually Happened

It was 93° out, the humidity the same.
He called at noon, with a couple quite lame.

I walked into the barn in the sweltering heat,
There stood six cows, all with bad feet.

"It's too hot for me to work, Doc," sweat on his
    nose,
"So I thought maybe you could just trim up their
    toes."

He sat back on a bale with a great big smile
And irritated me as he cracked open a Heileman's
    Old Style.

I did a good job as I trimmed up their toes;
One had foot rot, I could tell by my nose.

Glued a block on one, the heal was so rotten;
Another one got wrapped with gauze, iodine and
    cotton.

One was so bad she'll never walk on it, I guess.
One squirted pus when I opened a sole abscess.

An hour later, drenched with sweat and manure,
I was totally drained as I walked out the barn door.

"I can't pay you today, the checkbook has no money,
    I fear.
You go ahead and clean up, I'm going to town for
    more beer."

187

Ahh…Why don't we just give her a little more time to see if she is pregnant?

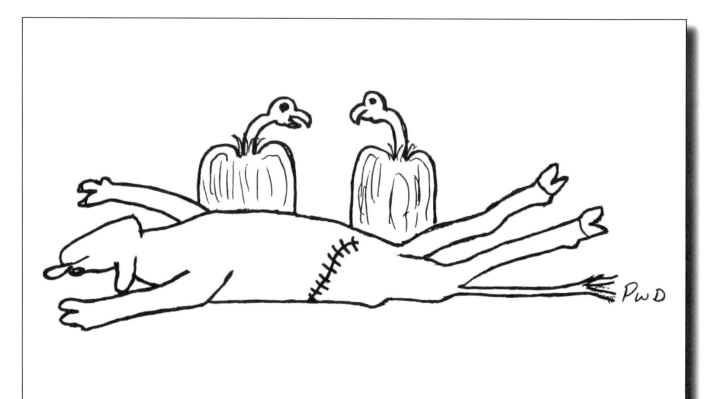

I just hate these dead surgeries. All those darn stitches.

A Day In September

I woke with the sun and to a milk fever I raced.
While I was there I looked at a fresh heifer that was
  displaced.

I then palpated 18 for pregnancy and uterus check;
While I wasn't looking, one wrapped its tail around
  my neck.

I went up a valley with the oak trees turning a
  beautiful magenta,
At that farm I checked a foot and removed a
  placenta.

I had three calls close together, west of me,
But I had to go north and deliver an OB.

While up north here, just down the road a mile and
  a half,
Was another call for one off feed and a poor doing
  calf.

Then south I went, one had pushed out her womb;
Next I saw a cow full of acorns, my prognosis was
  doom.

Finally out west I did vaccinate and infuse one
  draining.
Then a toxic mastitis, the milk was so thick they had
  trouble straining.

Back to the DA heifer for surgery, the milk fever was
  up.
It's a little after six; I'll be late for sup.

"If she's really sick, yes, I'll see her yet tonight."
But, I'll stop by my place first and grab a quick bite.

Now it's eight p.m., I'm caught up, to home I will
  race;
Clean up my vehicle, gas up and medicine to
  replace!

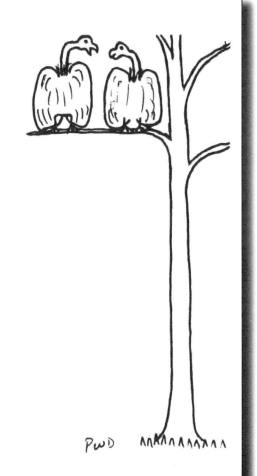

Have you noticed a change in the crows' hunting habits since we invaded their area?

Do you always order your meat rare?

Please, please…no more hop polkas or I'll turn to butter.

Displaced Abomasum

I was checking a beautiful Holstein for Walter
   Snead;
She was fresh ten days and wouldn't eat feed.

"I know she cleaned, and at first milked a lot.
This morning when I milked her only eight pounds I
   got.

Yes, Doc, she did have diarrhea, her stool was real
   loose;
She won't get up, lays like an old moose.

The only thing she will do is chew on hay some.
Oh, no, Doc, not a left-side displaced abomasum."

"Sell her or do surgery, if you want her around.
This way she'll never eat the feed you have ground."

"Go get your equipment and bring in your knife.
She's one of the best cows I've had in my life."

We shaved and we froze and cleaned up her left side.
We left her standing, her nose it was tied.

It wasn't too long, we had her abomasum back;
He helped me by putting on a tail-jack.

"Start her slow on grain, and give her some hay;
I'll stop in and see her, the afternoon of Thursday."

"Hey, Doc! She's eating her feed and filling the pail.
I just sent your check, you should get it in the mail."

In the winter this happens about every other day.
You walk into the barn and see that left-side D.A.

CARTER

Your place or mine?

191

Close Call

One early morning, I got a milk fever call south of town about 12 miles. As I'm dreamily driving along, I pass by Marvin's place and look up and see fire coming out of a wooden cupola and the entire roof is smoking.

The barn had been converted from a stanchion barn for about 80 cows into a parlor system. The parlor for milking was on the north side next to the holding area where the cows are gathered before milking. The remainder of the building is loose housing with a bedding pack open to the south.

I then see Marvin running frantically out of a side door in the holding area where the cows never exit back into the parlor. I back up, pull into his yard, and park down by the road to be away from the barn. In the back of my mind, I know I have a milk fever waiting so will need to get my pickup out.

I run up to the barn, and Marvin, who is normally a calm, logical guy, yells to me, "The cows are all in the holding area. The parlor is too hot to run the cows through, they won't go out the side door as I've never used it before, and they refuse to go back into the bedding pack. They are all going to burn, the stupid idiots!"

I grabbed a 2x4 that was by the barn entry and said, "I'm a stranger. Maybe I can help."

We went into the holding area. I started slapping cows in the head to turn them around, yelling, slapping, creating mass chaos. He did the same.

Finally, after about 5 minutes of our antics, one old lead cow headed south. With a lot of persuasion, we had about half of them out into the cow yard. We were about 40% of the way into the barn from the south side when the haymow floor falls in on the north side by the parlor. We were hit on our backs by a rush of smoke and hot, hot air. The rest of the cows felt this and

we all ran out together. That lead cow came around and wanted back in.

I yelled, " Marvin, open the cow yard gate and let's get them out of here."

We had one heck of a time to keep that lead cow from going back into that barn. The barn was a total loss. Only a few calves died in the fire, the majority of the animals got out.

The next day, I got up and had a sore throat and a deep cough. I had inhaled too much smoke and heat for my lungs.

The cows were moved to an empty barn and commenced to start with pneumonia as their lungs filled up with fluid. The insurance company decided that they should cash out by sending them to slaughter before tragedy set in. Three or four died before they got to slaughter. I hadn't been in the heat as long as the cows so I survived without any treatment and escaped slaughter, as did Marvin.

192

Damn. I always hated colics. Now this.

Well, yes, it works, but there have been some modern methods
of treating bloat developed recently.

Thanks for the delicious fresh soup. Say, how's that dog doing that I treated last week?

The Snare

I had a strange call from over by Blair,
Seems this calf had himself caught in a snare.

"Strange problem you've got, Ole, for this here rural
   region."
You see, my client was a full-blooded Norwegian.

"I've seen cattle in trees, trout streams and in mud
   they did mire,
But, never have I fixed one caught in a snare made
   of wire."

"Well, you see, Doc, I've got this electric fence that
   the deer keep breaking.
Because it's the season of rut, they've no respect
   with their lovemaking.

So, I got me a bow-hunting license, thought I'd pay
   to be fair,
Then I'd catch me a buck with this doe scented snare.

But, I guess the DNR will have the last laugh,
I'm out my license money, no venison and a bill on
   the calf."

194

I'm sorry, Mr. Bunyan, but all systems are normal. I have no idea why your ox is blue.

Just a couple of skin bleeders.

Best darn cowhand I've got.

Lucky thing you didn't fall off, Tonto.

Doc's the only guy I know who shoots only does.

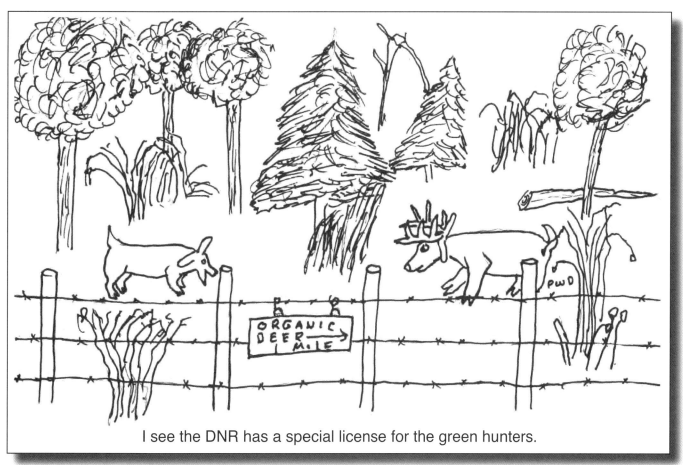

I see the DNR has a special license for the green hunters.

Extra Teat

When heifer calves get vaccinated for brucellosis and dehorned, there are always a few animals that would have an extra teat or sometimes two. These are usually located in an inopportune spot on the udder, so that when the animal is mature it will interfere with the milking machine.

I had a client with a registered herd and he had very good production. His bull calves were in demand because of his production and blood lines.

He sold his farm, with the majority of the cattle, to his son. He had one heifer that he kept ownership of to sell at a national auction, as he expected big money for its genetics.

Well, his son and I were going through the pen of heifers, vaccinating and dehorning, and he mentioned his father's super calf was in the pen and he pointed her out to me. When we got to her I noticed she had an extra teat and I mentioned it and he said I should be sure that we removed it when we got to her. Now this extra teat was right in between the two front teats, not in back like the majority of the extras. As I was getting ready to remove the extra teat, I noticed out of the corner of my eye, his father was entering the barn and was making his way towards us. Not really thinking things through, as I knew this fellow was Joe Serious, a little negative, without a sense of humor; I cut the teat off, held it up with my scissors and said to his son, "Darn, I cut the wrong one off."

Over the manger his dad flew – didn't even use the gate – and says, somewhat hysterical, "No, Doc, no you didn't!"

I looked at him, smiled and said, "Just kidding!"

He didn't think it was a darn bit funny and commenced to curse me out for doing something so mean and stupid.

Meanwhile, his son is laughing his head off, making it worse.

Never, never josh a humorless person.

Look Out, Deer!

Hunting the Whitetail Deer is a national holiday in Wisconsin. (My boys and I actually think it's more important than, say, Thanksgiving or Christmas!!) With our hilly wood-covered land and farms spread all over the hills, it is perfect habitat for deer. Over 700,000 orange clothed hunters hit the woods on opening day.

In all my practice years, with all my driving, I hit four deer with the front of my vehicle, killing three instantly and watching one slide up my hood, sliding off the side and running away. I also had two run right into the side of my pickup, killing themselves. I also had lots of near misses. When driving, you are always looking out of the corner of your eye for deer eyes or movement in the fields. Some fields were feeding areas, so that spawned a whole crowd of road-trippers with spotlights shining deer.

One fall, my nephew, Craig, who was a recent graduate from high school in Hawaii, came to stay with us. He got the deer hunting fever, like everyone else in Wisconsin. I had a good friend from the Twin Cities, in Minnesota, who had an excellent wooded farm with a neat old log cabin on it where he and three of his buddies would come down to every year to hunt. I would go out on Friday evening before the hunt and visit with them, so naturally that fall my nephew went along with me. About two miles before his hunting land, there was a hayfield on the left that was a known, well frequented, feeding area for deer. I came around a little curve and sure enough, my keen deer eye saw many silhouettes and eyes on the left – like a whole lot of them.

"Wow!" I said, "see the deer!"

I was focused on the left, a whole herd of deer. Meanwhile, right in front of me is a 1½ year old 7-point buck running slowly on the very edge of the road, on the shoulder. My nephew is watching that deer and thinks that's what I was talking about. We're going about 35 miles per hour when that buck decides to cross the road right in front of me, as I continue watching the herd on the left. Smack! I whack that little buck with my plastic radiator grill. I knock its feet right out from under it and it slides down the road on its side. Thankfully I was going slow enough not to hurt the deer seriously. It got up and ran up into the field where it belonged. My grill was 700 pieces of black plastic on the road. Didn't hurt the radiator or headlights. It just took out the grill.

My nephew said, "Why didn't you stop, Uncle Paul? You knew that deer was going to cross the road."

"I never even saw that one little buck. I was looking at that huge herd of deer in the field."

"What deer?" he asked.

I replied, "That's the problem with you young folks, you just don't look!"

Hey, Strep, there's a new dose of penicillin in the kidneys.
No, Staph, my hubby and I are headed to the liver. A new batch of organo-phosphates just got deposited and we don't want to miss out on OP.

Winter

It's January the eighth, and twenty-two below;
As you head to the garage, you hope the vehicle will
  go.

The engine starts the first spin of the crank,
Your first call is a mastitis at Wilber Frank.

You walk in the barn; Wilber's face couldn't be redder.
He just broke the apron chain on his frozen manure
  spreader.

"Doc, do you know how cold it is out laying on your
  back,
Undoing a broken chain with manure dripping
  through a crack?"

The next call, a uterus infusion at the farm of Roland
  Rup.
Found Roland covered with haylage, as his silo
  unloader was froze up.

"Doc, do you know how dirty, dark and cold it is in
  a silo chute?
My cows are all hungry and nervous. That one's
  mean as a brute."

My next call was an off feed at Anthony Klop.
He lives west of town on the county blacktop.

"How come you're not milking yet, Tony?" my
  question was a depressor.
"My milk house is all froze up," he snarled,
  "including my compressor."

I gave his cow a physical from her nose to her feet,
While he was in the milk house, thawing out things
  with some heat.

I hopped in my vehicle and speedily I did race
To put in a prolapse over at the Nelson place.

"I'll need lots of hot water to put this back in."
I could tell I said something wrong as he lowered his
  chin.

"There isn't a pipe that's not froze up tight as a
  drum.
I'll hop in the pickup, go over to the neighbor and
  get some."

The next few calls went without a hitch;
There was one half-ton pickup stalled in a ditch.

The weatherman on the radio said tonight will be
  thirty below.
Of all the farmers there won't be a happy one, I
  know.

I head home in the cold and put my vehicle in the
  shed,
And thank goodness I fix cows, not farms, as I crawl
  into bed!

Bacteria

What would it be like if Louie Pasteur
Hadn't found those little bugs that grow in manure?

There would be no bacteria; they wouldn't exist;
We wouldn't use thermometers we shake with our
   wrist.

There would be no bacterins to protect from disease;
We wouldn't even know the reason we sneeze.

Now we've got vaccines, killed and modified live;
And nearly everybody vaccinates annually with
   Lepto-Five.

We vaccinate in the rump, nasally and mouth;
All creatures get sick, even down south.

Now, they have isolated many different strains;
One from the east and two from the plains.

We sell and we vaccinate on into the night;
The cost of these new vaccines is really a fright.

But they still get sick, in stanchions and free stalls;
And, why in the world have I got so many calls.

One can never be sure, but I think it died from Blackleg.

I told you he'd get that girl in trouble.

My mother says I shouldn't go out with you wild boys from the ears.

I'm leaving you. My attorney will be in touch.
By the way, you have 10,000 nits to support.

Electronics

My grandfather was of the Horse Era, and never
   mastered the combustion engine of the tractor.
He would flood them, get them stuck, run thru fences
   – his horsemanship was a negative factor.

I'm faced, in this generation, with a dilemma of a
   different magnitude – I just freeze.
That's the electronic gadgetry that runs our world
   that we import from the Japanese.

I can't turn on our Tandy computer; I just stand
   there and gape.
I can't turn off our VCR after 'Dallas' without
   ruining the tape.

With this Electronic Age, I'll admit I'm a real mess;
My nine year-old can run Guppy mouths around,
   until everything's ate up, I guess.

I played baseball when young, I was pretty good at
   third base.
But to hit that ball on that little electronic baseball
   game gets me red in the face.

We got a toy electronic keyboard with eighteen
   buttons for pre-school boys.
It's got rhythm and blues; the melodies coming out
   are beautiful noise.

They can play Rumba, Bluegrass, the Cha-cha and
   two-step;
I can press a few buttons and turn on a switch, not
   one little blep.

We went and had pizza, with these electronic games,
   and saw a singing bear.
I wanted to drive that electronic car, but totaled it
   out before I sat in the chair.

I've resigned myself to being a real electronic sap;
I can now slowly see the width of the 'Generation
   Gap!'

Farmers

There are three types of farmers.

After watching how dairy farms are managed, one notices a pattern. You see, to be a good veterinarian, one has to be a good observer. If you don't observe everything and everybody, you will not be a good veterinarian.

The first type is the Machinery Man. A born mechanic, he usually has a newer line of equipment. It's all housed, quite clean and he knows every detail - width, horse power, what it's capable of doing, on and on…

He is mighty proud of his machinery.

The second type if the Cow Man. He knows every cow, their mother, their daughters, when they were born, how much butter fat they have, and if she is ornery, calm or light quartered. He knows who he is going to breed her to for stronger legs or calving ease or higher butter fat. He spends all his extra time in the barn with his cows. He's in sync with them, adores them, talks to them and treats them very humanely. They are his passion.

His main tractor is some old, beat up 35 year old, 50 HP Massey Ferguson with one fender off. It hardly has any paint on and has a seat on it, from who knows where, with a worn out cushion, and it needs a muffler. The tires don't match, but it runs and that's all he cares about. He's got a Farmall H, that is in even worse shape, for a second tractor and an old "A" John Deere in the box elders that hasn't run for at least 10 years. The rest of his machinery is in the same shape. Before he can bale a few square bales, he's got to spend two days fixing things, hunting up parts, getting air in the tires and cleaning the mouse nests out of where the twine spool goes. In fact, the last two years he has hired a custom operator to put up the big squares as all his kids have graduated high school and gone to college. Neither boy is coming back to farm and his daughter, who used to drive the baler, is an RN, married with two children living 150 miles away. But guess what? He loves his cows!

The third type farmer, and unfortunately these are the fewest, are the Soils Men. They love their dirt!

In the organic/sustainable/biological, these are the best farmers as everything starts in the soil. They do soil analysis, watch their calcium levels and know who William Albrecht and Carey Reams are. Foliar feeding the plant is the next big step they are just entering into. They know what good chocolate cake soil is like. They don't want to hurt their earth worms and are improving their soil and ecosystem. They can connect: healthy soil + healthy plant + healthy cow = healthy food, which means healthy people.

Organic/sustainable farming started out as a dream of a few people a few decades ago and fortunately, it's growing by leaps and bounds in the United States in spite

No, there's no written test involved when we have a dairy assignment.

of our great bio-tech world. The CSA's and farmer's markets are all about caring for the soil. This group has minimal machinery and equipment and have a good feeling for plants and livestock.

Now, the trick is to get all three types of farmers rolled into one person. A good machinery man, a good animal man and a good soils man. When you have that you will see a fine tuned operation that is flat-out turning a profit.

If I were to take 1,000 farmers in dairying today, I would expect to see 40% cow men, 40% machinery men and 20% soils men. About 35% of the soils men would be good in two fields and 5-10% would excel in all three areas.

I once had one of the best farmers I've ever worked with that was excellent in all three tell me his philosophy: little things don't mean much, they mean it all.

When I get on farms as a con-

sultant, I'm categorizing each and every one into the three groups and to what degree. In partnerships of two and three people, it's important to recognize each one's abilities and to put them in the correct slot. You don't want a good cow man taking care of machinery! When hiring help on your farm or company, keep this in mind: fit the man to the job!

Yep, she's cycling.

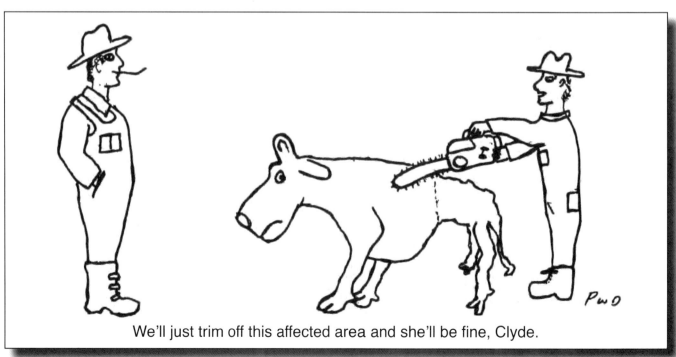

We'll just trim off this affected area and she'll be fine, Clyde.

I think I'm getting a head ache.

Been losing a few lately, Doc?

Cattle Dealer

I had a good account that I got to know early in my career who was a cattle dealer. Back in the 60s and early 70s, every cow he bought and sold had to have a negative blood test for Brucellosis. This meant that I made 3-4 calls a week to bleed his recently purchased cows, usually at night. He always had a good Ole and Lena joke and I treated him lightly on my fee. We became good friends and I respected his honesty.

Well, about that time, another disease problem became quite common, called Johnes. This was a chronic diarrhea that was caused by a bacteria invading the small intestines lining, thickening it, causing a green pea soup diarrhea. They get a wet, rat like tail and lose weight. As soon as a farmer would detect one, they went to the sale barn as there was no treatment and once they started showing clinical signs, they went down hill fast.

Well one night this black Holstein appeared in my cattle dealer client's barn. As I drew blood, I noticed she was a little thin, but she looked like a good cow that could produce and had maybe milked herself a little thin. I had a good client about 15 miles east of me that called me about 10-12 days later for some miscellaneous veterinary work. As I was there, he asked me to take a look at this black Holstein cow that he had just purchased from a cattle dealer. She looked fine when he purchased her but had a nasty diarrhea.

I checked her over, and sure enough, here was the cow I had checked less than two weeks ago at my cattle dealer friend. She had blown up into a classic Johnes. Wet, ratty tail, skin under the jaw (bottle jaw) hanging down, getting quite thin. I told him to call the cattle dealer back as he usually stood behind his sales.

So, he goes to the phone and calls the dealer up and tells him that the black cow he had bought blew up with Johnes. My dealer buddy tells the farmer to put her on hay. Fill her up with dry hay and take her off grain. He said to wash and shampoo her tail. Clean her up good and take her to the sale barn and sell her the next day. He said he would pay the farmer the difference that he ended up short. Done deal.

Six days later, I make my evening journey out to my cattle dealers to draw my bloods from his new acquisitions, and here stands four cows. One is black and spits out green pea soup diarrhea all over her nice tail. I chuckle as the dealer walks in and I tell him he should call and thank my farmer client for the fine job of cleaning up that Johnes cow. He says, "Why should I do that? I say, "Cause she looked so good you just bought her back!"

I take it you don't think her odds are too good, Doc?

I just hate it when he says "trust me."

Farm Charity Bazaar

Cow 1: I'll donate butter. Cow 2: I'll donate milk. Cow 3: I'll donate cheese.

I suspect she's been eating a few too many acorns, Clyde.

Wildlife

I put on many miles chasing sick cows day and night.
I truly enjoy all the wildlife; it's such a delight.

I look for birds and animals, just out of habit.
Look, over there, sits a Cottontail Rabbit.

Whether it's sunny and bright or dark and murky,
Just around the bend we might see a Wild Turkey.

The birds are aplenty; they're all over, never fail;
From Bald Eagles, Doves, Wood Peckers and a
   Hawk with a red tail.

The neatest sight, whether it's at dusk or dawn,
Is to look up and see a Doe with a Fawn.

The chirpy little Squirrels, a scurrying away;
They can really move, no matter if they're red or gray.

We don't have Caribou, Elk or Mountain Goat,
But lately we're sure seeing some wild Coyote.

The sleek Red Fox is always standing and looking
   back,
To make sure there isn't some farm dog on his track.

The plumpest animal that I've ever seen,
Is the flat tail Beaver, that wood eating machine.

The reptile that makes women scream at a glance,
The big Bull Snake on the road, doesn't have a chance.

In the fall all the autumn colors can be real pleasant,
Especially when you see a long-tailed rooster
   Pheasant.

The only one I harvest, so slow and so dapper,
Is for our soup, a great big old Snapper.

They're on the bottom roads in the spring; they've
   lost their winter fat;
Choosing a slough to live in with the Bullfrogs and
   Muskrat.

Only in the spring does everything flower and
   blossom.
Twelve months of the year, slowly trudging along is
   the primitive Opossum.

The last guy you want to run over on a township or
   county trunk,
Is the black and white, waddling, smelly old Skunk.

So, I'll just keep on trucking the miles away,
And keep on looking in the hills and valleys for more
   Wildlife today.

208

That's not what they mean when they say to
"challenge feed high producers," Clyde.

The Coat

For years when I was collecting cast iron, I would go to a big swap meet in Waukee, Iowa, west of Des Moines. There was a fellow there that sold used coats and other items in the flea market area. I spotted a red and black flannel jacket that was in good shape, fit me really good and was only a few bucks, so I bought it. It looked like the type of jacket a logger would wear. Well, my wife absolutely hated it.

You see, early in our marriage, we went to a movie by the name of "Deranged." It was the story of Ed Gien from Plain, Wisconsin. Ed Gien had an odd habit of robbing graves and doing a lot of strange things with the fruits of his labor. This happened in the 50s in Central Wisconsin. My wife and sister-in-law decided they had to see this movie but were scared to go alone,

so I was elected to accompany them; and, of course, had to sit in the middle. I had written a poem about Ed Gien when I was in high school, for an English assignment, so I knew exactly how twisted Ed Gien was and how scary the movie was apt to be. My wife and her sister, being somewhat younger, knew very little of his history. I felt quite bruised and beaten up by the time we left the theatre, as the movie was very scary and a real gripper. The main character, portraying Ed Gien, wore a red and black plaid jacket, a real dandy, identical to the jacket I had just bought.

This jacket was perfect for veterinary calls. It fit nice and I liked my red-necked, tough logger image. An added plus, was I knew my wife hated it every time I wore it! One day I

did a stomach surgery in a barn on a fairly mild day, and I drove away from the farm forgetting my jacket. Not remembering where I left it, I figured, 'Oh well, I got good mileage out of what little it cost me, and my wife was happier, so….so be it.'

The next winter I return to this barn and here I spot my jacket on the back of my farm client. I could tell it was mine as the pocket had a little rip in it, and sure enough, so did the one he had on. Not wanting to cause a fuss or embarrass him, I said, "Hey, nice jacket. I like those colors. Where'd you pick up a find like that?"

"No idea," he said. "One day it just showed up. It fit me and I've always liked black and red. Can't take it in the house though, my wife hates the thing."

So be it.

How many test tube babies have you had?

Molu Calves

Many young calves, after they are weaned, get moved on into new housing to make room for the newly born.

Where to go with them is a question on many older, smaller homesteads. A lot of these farms were diversified back in the 30s and 40s and had chicken coops for 300 to 500 laying hens. The wife was most always in charge of the chickens. The egg money was a very nice addition to the household expenses. It bought the groceries, the kids' school clothes, lunch money and many other incidentals.

In the 50s and 60s, the chicken industry died and went to large scale operations. So, the chicken house, still being a useable building, had the roosts and the laying boxes for eggs thrown out, and by putting in a couple of pens, we had a calf barn.

Most chicken coops had ventilation with a fairly big wall fan for temperature control and to keep the air from getting strong.

Calves, by nature, are curious creatures and investigate everything in their environment. The fan, with its noise and motion, attracts their attention.

How do they check things out with their sense of smell and taste?

Now, a bovine, because it uses its tongue as a prehensile organ in eating grass, can stick their tongue out a very long ways.

So, this fan with sharp blades is running at very high RPM and this calf sticks his tongue into it and CHOP. Most of the time, one to two inches gets cut slick and clean off. Sometimes, it will get cut two thirds of the way through, and you will have a calf with a very sore mouth. It has difficulty eating, does a lot of salivating and is uncomfortable. Those you cut off the rest of the way.

A sore tongued calf will get thin and obviously will go backward.

I would recommend to put them back on milk or try giving them a wetter diet while the tongue healed up. There was usually not a lot of blood, as the elastic arteries would pull back into the tongue muscle.

I would also recommend putting up some shielding to keep the tongue out of the fan. You could always tell the short tongued cows later in life as they had a lisp in their speech and when they bellered "moo," it always came out "molu."

Doctor, would you cut this growth off Junior? It's just in the way.

So, how many times did you give him 20cc?

OK, Quincy, mark this down. Two Firestone, one B.F.Goodrich and one Michelin.

All Junior wants to do is walk around on two feet and wave his arms.
Can you check out his back, Doctor?

Doctor, could you cut off the extra finger on Chippy's left hand? It looks terrible.

I see they have GMO'd corn with hickory trees so our bacon will taste like it is hickory smoked.

Did you hear the one about Ole and Lena?

So, Clyde, how's this night time grazing going for your herd?

Man, am I hungry.
You didn't see which way the vet went this morning?

Been having a tough day, Doc?

We Are On Top

Years ago when life first evolved;
Let me tell you how it could have resolved.

There once was a flea that came up lame;
He was in dire straits in the feeding game.

It hurt so bad, he had a bad limp;
His buddies laughed, and called him "The Gimp."

He got thin, debilitated and was losing his will;
Until he visited his uncles, Wilber and Orville.

In the flea circles, their brains were a genius thing;
So they sat down and invented an efficient wing.

You just strap these things on your side, here;
Then use the air currents as a glider.

The flea became famous, and among other things;
His multitude of offspring all had wings.

The one offspring, with a gleam in his eye;
Said "I'm not gliding, I'll just move them and fly."

A whole new genre was then on the scene;
Those pesky flies, that land on the screen.

The fleas and the flies, then called it cahoots;
Let's settle this young planet and put down our roots.

We'll sit here and think, in the sun and the rain;
And become creatures here, with the largest brain.

You see, to come out on top of the evolutionary
    season;
To conquer other species, they had to know how to
    reason.

So the two species built cities, without friction or
    question;
They invented the engine, of internal combustion.

The homo sapien species were nearly last in the
    race;
As they had a very small cranial space.

We remained a predator, and followed wandering
    courses;
They controlled us by their Department of Natural
    Resources.

So you can see we could be on a totally different
    time train,
We are only where we are at because of our brain.

The Thrill

As I've gotten older, less hyper, and actually mellow;
I still have delight when I see Spring's first pussy
    willow.

When a son, after Hunter Safety, shoots his first
    deer;
What a thrill to see his face, and the story you'll
    hear.

When a daughter has a recital, and never misses a
    note;
You sat there so stoic and confident, with sweat
    under your coat.

When a son wrestles in a tourney, and takes first
    place;
You're happy for him, he can tell by your face.

When a client says "Thanks, Doc, for coming in the
    middle of the night;"
You get a warm feeling, knowing you've done right.

The thrill of success, when the bills are paid;
Dividing the fruits from the money you've made.

When you go to bed when a tough day is through;
And your wife quietly says "I love you."

Your heart lifts its beat, your bosom is filled;
Life is so sweet, your body is thrilled.

The times you worried, the parental dedication;
Is all rewarded at the high school or college
    graduation.

There are losses, at times you encounter defeat;
It makes one stronger and better, when success won't
    repeat.

While you are able and conscious, in the valleys and
    hills;
Always take a second to cherish your thrills.

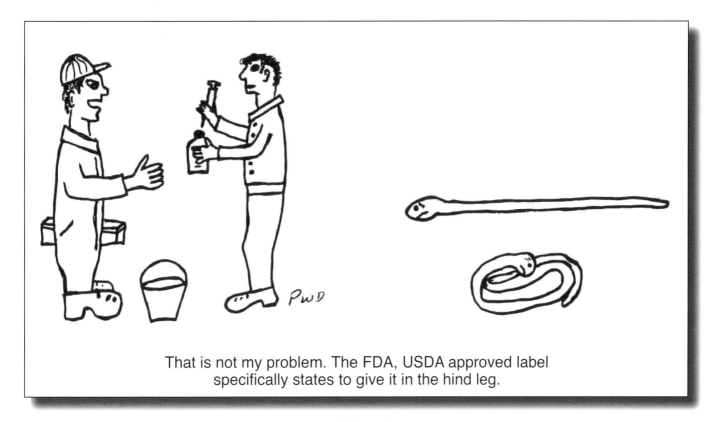

That is not my problem. The FDA, USDA approved label
specifically states to give it in the hind leg.

Dad

When I was eight, Pappy gave me a baseball glove
  signed by Ty Cobb;
He taught me to hit, pick up grounders, and second
  to rob.

He'd discipline me sometimes – my behavior caused
  a rage;
When I was twelve, he gave me a pump twenty gauge.

Weather would be sunny, nice and pleasant;
We'd be out with Dad, getting our limit of pheasant.

Baling hay was done, all that sweating was hard on
  your liver;
"Let's go, boys, we'll do some fishing on the
  Mississippi River."

We would set up a tent, take juice and sandwiches of
  ham;
And fish off the rock pile, over by Whitman Dam.

Football season, I was a tackle, we were in first place;
There on the sidelines, in the cold rain, stood Dad
  with a smile on his face.

Baseball season, farmers planting oats and making
  crop plans;
Always, there sat Pops, in the bleachers with the rest
  of the fans.

We never had much money, but that wasn't so bad;
For moral support and help, there was my Dad.

College came, there were no scholarships or grants
  to take;
Just my Dad, with some words to the wise and a
  handshake.

When I got out of college, went into business and
  started to toil;
There was my Dad coming over, reminding me to
  change that oil.

And when one gets older, kids of your own all on the
  go;
I can now understand why, occasionally, Dad said
  "No."

I get sort of down and sometimes depressed;
Because it's only on two generations that memories
  get impressed.

As I'm going back home tonight, a tear I cried;
Because today, my good old Dad died.
                                    (October 24, 1989)

Upbringing

My dad was a grass roots, German type guy;
When he said "jump," we said "how high."

A hand shake means a deal is a deal;
No contract, documents or attorney's spiel.

His book learning was minor, he didn't frequent any
   schools;
Common sense and logic, were his educational tools.

With math, figures and numbers, no calculator,
   instead,
He multiplied and divided it all in his head.

When purchasing a car, from some salesman with a
   smile;
He knew what he'd pay, he'd figure his cost per
   mile.

He said education is good, learn all about muscle
   and bone;
But always know how to think on your own.

When other guys drift and stray in the night;
Don't lose your principles; know wrong from right.

When I'm explaining to my offspring what's naughty
   and nice;
I find myself using my father's same advice.

I'll just keep passing on all my acquired and
   inherited upbringing;
So I can leave the next generation with their ears
   ringing.

It's gotta be the shoes.

Bet my dad can lick your dad.

They taste better everyday lately. There's less hair on them.

A Million Miles

I've driven over a million miles on our country's
  back roads;
I've accidentally hit deer, pancaked opossums and
  squashed many toads.

I slow down and swerve to miss Mother Nature's
  wild critters;
In the fall when the deer are in rut, I drive with the
  jitters.

I keep yearly statistics, it gives driving a fun twist;
Every road kill, posthumously, gets recorded on my
  Dead Animal List.

Squirrels are down, rabbits are up; last week I
  recorded a dead mink.
Signs don't help, I just wish those poor animals
  could think.

I observe the crops, weed control, who's cutting
  second crop hay;
There's a logger cutting red oak veneer and hauling
  them away.

I just sit and observe the happenings in the quiet
  rural land;
Yet keeping in touch with the world with my radio's
  FM band.

I'm up-to-date on Noriega, Gorbachev, the Balance
  of Trade and Consumers Price Index.
The news is negative and depressing. With our good
  times, they make it sound like we're in a fix.

I'm an optimist, though, so I just turn off the bad
  news with a switch;
And, I'll keep on pursuing my goals in my little
  country niche!

I Won

Now that you've finished this book, on me your
  conclusion is made;
You've rendered a judgment, I've been given a grade.

My grammar is bad, I'm colloquial, I even use ain't;
One thing for certain, my wife is a saint.

She lives with this weird vet, I'm not in the fast lane;
My humor is subtle, I've got an unusual thinking
  brain.

I hope I've lightened your spirit, and gave you a
  smile;
I worked years on these poems, yes, it took quite a
  while.

But don't pity me, I'm very happy, don't laugh;
Whatever you paid for this book, my profit is half!

I just love his humor, but hate his cold arm.

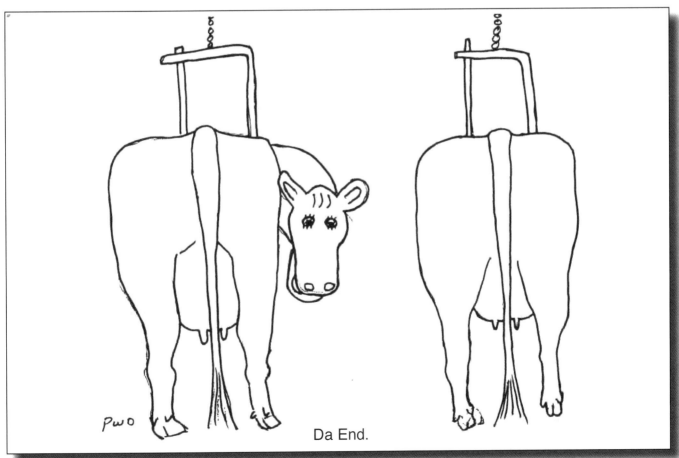

Da End.

Dr. Paul W. Dettloff, DVM
Biological Sustainable Veterinarian

Dettloff
Tree Farm

Minnesota